For Sue and Eloise.
In hopes of a better world for
David, Jason, Sam, Melissa, and Will.

Preface

Generations of Air: What Will We Leave Our Children?

~~~~~~~~~~~~~~~~~~~~~~~~~~~~~~~~~~~~~~~

The composition of the troposphere, the lowest part of the atmosphere that supports all life on Earth, began to change significantly around the beginning of the century. In more ways than one, our study begins at the same time. In one sense, this book is about the atmosphere. In another sense, it is about life, our lives, the lives of the four generations of families that have spanned this time frame.

In 1906, our grandparents immigrated to the United States from Poland to seek a better life. First came William, who when arriving in this country took the name of the city in Poland from which he came— Kalisz, which was anglicized to Kalish. Then came his wife, Nettie. The courage of these immigrants to set out for an unknown land where the language and customs differed from their own can only be imagined from the perspective of our own rather sheltered and safe lives. They were driven by a will to better themselves and, more importantly, to insure that their children and their children's children would not have to suffer the hardships they had.

The miracle of America then was that it worked. Willie and Nettie's children all finished high school. They all lived in houses with indoor plumbing and heating. Our parents were truly living the "good life" compared to their immigrant parents. But they, too, suffered. There was the depression of the 1930s, and then World War II. But the dream of America has always been that the quality of life would continue to improve as each generation comes along. Our children's lives would somehow be better than ours. The children of these first-generation Americans, those of us who grew up in the post-World War II era, enjoyed an unprecedented life style. Energy was unlimited, the natural environment was bountiful. We had automobiles, washing machines and dryers, and television. The future seemed to stretch in an unbroken highway of "good things for good people," of "better living through chemistry." The road belonged to America. Had Grandpa Willie lived longer, he would have truly been amazed at the lives of his grandchildren, at the sheer largesse his adopted country had bestowed upon his heirs.

But there was a price. The composition of the atmosphere that was already beginning to change in Grandpa Willie's time altered drastically during the 1950s as America became a land of automobiles. Gasoline and oil were cheap, and the standard thinking of those days said we had enough to last us for centuries. There was no need to conserve, no need to practice restraint. Conservation meant going out to the woods to look at the trees and animals.

By the 1970s, concern for the environment grew

into a movement. The post-war generation, those entering adulthood in the 1960s, began to see the truth—there is no free lunch. By the 1970s the government began to pay attention and the Environmental Protection Agency was born. Some progress was made; our dirty air became perceptively cleaner. We no longer looked upon the internal combustion engine as a panacea, and began thinking of ways to conserve fuel, especially when the price of fuel skyrocketed in 1973 due to the Arab oil embargo.

But the 1980s saw us backslide. Our leadership seemed more concerned with reminding us of our heyday in the 1950s than in addressing the environmental problems that needed attention. Those who kept reminding us of such problems as acid rain and toxic metals, of our polluted waterways and air quality in our urban areas, were voices in the wilderness. We were told things were alright. We wanted to believe. We've paid a price for our naivete.

Back in the 1950s, life seemed so much less complicated because we knew so much less. We knew that fresh air was good, for example. Our parents would often tell us as they tucked us into bed at night that a day spent outdoors, playing hard, would bring a "good night's sleep." And we would still like to believe that now, with our own kids, but the truth is, things have changed. When we were kids, the air away from our large cities *was* good for us, it was cleaner than the air in the cities. But today, as we approach the 21st century, the composition of the air *all over* has changed. The change is subtle, hardly noticeable. Since the early 1970s the concentration of ozone in the lower atmosphere has increased at an average rate of between 1 and 2 percent per year. That is three to five times faster than the well-publicized increase of carbon dioxide.

But why is this increase in ozone so important? Because ozone is a poison. Because concentrations of ozone, even in many rural locations throughout the northern hemisphere, often reach levels capable of causing damage to human health. At current concentrations, ozone pollution costs the American farmers between 1 and 5 billion dollars. At current concentrations, the forests in Germany are beginning to die, and our own forests in the northeast and throughout the United States are showing signs of damage at higher elevations.

During the summer of 1988, one of the hottest on record, ozone concentrations were high enough on numerous occasions in the eastern third of the country to cause damage to young children's lungs. These levels affected not just children and adults in major cities, but also those in rural areas. This is quite a difference from the 1950s. Back then, ozone levels in the atmosphere only occasionally reached harmful levels, and then such incidents were limited to certain urban areas, like Los Angeles with its notorious smog.

Ozone is synthesized in the lower atmosphere through a series of chemical reactions. Three ingredients are needed to make large amounts of ozone efficiently: two gaseous precursors (a nonmethane hydrocarbon and an oxide of nitrogen), and sunlight. The largest source of both nonmethane hydrocarbons and oxides of nitrogen is automobile exhaust. At one time we erroneously believed that ozone pollution was actually going down during the 1970s. We believed that catalytic converters and the use of nonleaded gasoline would solve the problem for us since the converter promised to reduce hydrocarbon emissions. The EPA had promised that there would be no more dirty air by the year 1976. Then the deadline was extended a full 10 years by the Clean Air Act Amendments of 1977, which promised clean air by 1987. In addition, the government lowered the National Ambient Air Quality Standard (NAAQS) for ozone concentrations from .08 parts per million to .12 ppm. Was it any surprise that during the summer of 1988 ozone levels over most of the United States were the highest that had ever been measured? The end result is that virtually no progress has been made on a national level to abate widespread ozone pollution.

To be fair, one reason for our lack of progress is simply that we didn't completely understand the dynamics of ozone production. The atmospheric chemistry that we thought satisfactorily explained the formation of ozone pollution turned out to be more complicated than scientists in the 1960s thought. Also, we now know that ozone pollution is not a phenomenon confined exclusively to industrialized countries. The farmer in South America who burns his fields every year after the harvest is contributing to the ozone pollution problem just as much as the city commuter driving to work.

Despite the lack of action and interest during the past ten years, all is not lost. There are steps we can take to stop ozone pollution. To stop the trend of rising ozone levels in ambient air, we first must know what is happening, then take action. We *can* clean up the air. We can leave a legacy to our children of fresh air and hope, the promise that brought our grandparents to this country in the first place. That's the way Grandpa Willie would have wanted it, for his children, and for his children's children, all down the line.

Jack Fishman
Robert Kalish

Poquoson, Virginia
Bath, Maine

Acknowledgments

~~~~~~~~~~~~~~~~~~~~~~~~~~~~~~~~~~~~~~

This book would not have been possible without the help and encouragement of many others. We'd like to thank in particular Mike Snell, without whose initial enthusiasm and guidance this project would never have gotten off the ground; Dr. David McKee of the Environmental Protection Agency, Research Triangle Park, North Carolina, who provided us with much material used in this book; Dr. Robert Harriss of the University of New Hampshire for his constant encouragement; Dr. Barrett N. Rock, also of the University of New Hampshire, for sharing his research and providing a patient introduction to anatomical botany; Dr. Lamont Poole of NASA Langley Research Center, for his technical insight on PSCs; Dr. George Bokinsky of the Maine Medical Center in Portland, for his help in the chapter pertaining to human health effects of ozone pollution; the Maine Lung Association, for their help in tracking down published studies of ozone research; Linda Greenspan Regan, from Plenum Press, for her editorial guidance; and most of all we'd like to acknowledge the moral support of our mates, Sue and Eloise, without whose cooperation this task would have been so much more difficult.

Acknowledgments

Contents

~~~~~~

CHAPTER ONE

The Smoke-Filled Room

~~~~~~~~~~~~~~~~~~~~~~~~~~~~~~

The scene is a familiar bit of Americana—a summer camp nestled in the rolling mountains of central Pennsylvania. A horseshoe-shaped cluster of cabins hugs the shore of a pristine lake. Children are busy on the dock that juts out from the sandy beach. Canoes and small sailboats bob peacefully in the water. Alongside the cabins and dock is a grassy field, where a group of children are playing tag. Sam, an eight-year-old boy, is chasing his friend Melissa. The two of them are running as only children can—with a playful abandon reminiscent of puppies. They jump, they leap, they twist out of the way of the outstretched arm. Their bodies, so young and fresh, are running as hard as they can. Their lungs work as efficiently as machines—sucking in air, transfering oxygen to the blood in exchange for carbon dioxide.

But at this summer camp on this particular day something is different. Next to the playfield sits a trailer. Inside the trailer are shiny metal machines. At intervals during the day Melissa, Sam, and their friends stop by the trailer, where they are asked by a scientific team to breathe into the machines. Afterward, they are released and resume their activities.

Summer camp. The words evocative of mist-covered lakes, of campfires and singing, of evenings spent in pleasant fatigue after a day of "healthy, outdoor living." For years we've been sending our youngsters away to camp, away from the bustling, crowded cities with their bad air and noisy streets. For years we've assumed that a few weeks at camp, in the healthy environment of nature, served as a tonic to the child's mind and body. In some cities, a program to send ghetto children to camp is called the Fresh Air Fund.

But it may be time to rethink these assumptions. And the reason can be found inside the trailer Sam and Melissa and the rest of the campers visited regularly during their stay at camp.

In 1980 in rural Pennsylvania, and again in 1982 in rural New Jersey, two teams of scientists spent several weeks at two summer camps. Using modern equipment housed in trailers, the scientists regularly tested the lungs of campers after they had concluded various physically demanding activities. The result was alarming: healthy children exercising moderately on a warm, hazy day were damaging their lungs. Both camps were hundreds of miles away from any sources of air pollution and during the days of testing the air quality was well within the standards of "clean" air as set by the National Ambient Air Quality Standards (NAAQS).

The culprit in this tragic scene is ozone. Ozone is a pale blue gas found in minute quantities in the air at the earth's surface. Ozone is also poisonous to humans. For most of earth's history, ozone has been present in our air in such small quantities (much less than one part per million parts of air) that it did no harm. Other gases are also found in our air in small quantities, and these are called trace gases because just a "trace" of them exists along with the main components of air: oxygen, nitrogen, and water vapor. From a distance, ozone is a lifegiver. It is the ozone "shield" in the stratosphere that prevents harmful ultraviolet radiation from reaching the earth, allowing life as we know it to continue.

At the earth's surface, ozone is manufactured by electrical discharges interacting with oxygen. The smell often noticed after a thunderstorm is ozone, produced by lightning interacting with the atmosphere. Ozone is also manufactured artificially, present in electrical conductors and used as a cleansing agent in swimming pools and municipal water supplies. Surface ozone is also caused when certain combinations of gases in the atmosphere interact with sunlight. Ozone is the main villain in photochemical smog, meaning smog produced by the interaction of sunlight and certain components of the air, such as that which plagues the Los Angeles basin regularly. Historically, ozone has been present in our air in such minute quantities that it posed no danger to human or plant life. It has always been a part of the chemical makeup of our atmosphere, present in such minute quantities at the surface that it was considered a "trace gas," along with methane, carbon dioxide, and others.

What is so alarming about the findings with the summer campers is that the testing was done on days on which there was no apparent pollution. Yet the levels of ozone on "clean" days were still higher than they were 100 years ago. Even at levels deemed moderate and safe by the NAAQS, about .08 ppm (parts per million) or below, ozone is capable of damaging the lungs.

Ozone is a form of the element oxygen. But whereas oxygen molecules have two atoms (O_2), the ozone molecule has three oxygen atoms (O_3). To manufacture ozone, you simply have to add a free oxygen atom to a molecule of oxygen. Electromagnetic forces are perfect for separating oxygen atoms from other gases and allowing them to team up with available oxygen to form ozone. Hence the smell of ozone after a thunderstorm. In effect, this is what happens to your spark plug wires in your automobile. The electric current running through the wire leaks out, despite the insulation around the wire. This creates ozone in the thin layer of air adjacent to the outside of the wire. The ozone eventually destroys the rubber insulation, causing either a short circuit or the engine's gradual loss of power.

In the atmosphere the process is dependent on sunlight. At the upper reaches of the earth's atmosphere, above about 8 miles, begins a layer called the stratosphere. It is here that 90 percent of the earth's ozone resides. The air in the stratosphere is much thinner than that at the surface, comprising only about 15 percent of the atmosphere's mass. This lack of density appeals to photons, parcels of electromagnetic energy, most of which are put out by the sun. They possess no mass but supply the means by which light's energy reaches the atmosphere. Some of these photons contain enough energy to break up the two-atom oxygen molecules into free-ranging oxygen atoms. It becomes quite busy in the stratosphere with all those oxygen molecules breaking apart and their free-running atoms then teaming up with other oxygen molecules to make the triatomic (three-atoms) form of oxygen, which is ozone.

This process occurs all the time. The ozone created up in the stratosphere is called the *ozone shield*, which is chiefly responsible for permitting life as we know it to exist here on the surface of the earth. Without the ozone interacting with the high-energy photons of the ultraviolet portion of the sun's spectrum, those same photons would reach the earth, bombarding us with more of the harmful ultraviolet rays of the sun's spectrum. If these harmful ultraviolet rays were allowed to penetrate to the surface, many of the life forms on the planet would not survive since the delicate balance that supports such life would be destroyed.

The process that makes ozone on the surface is similar to the one producing it in the stratosphere, with one important difference. At the surface there aren't enough high-energy photons to break down the oxygen molecule. Still, there are other gases that can be used to create ozone. One of them is nitrogen dioxide (NO_2). NO_2 has an affinity for the less energetic photons in the visible portion of the electromagnetic spectrum (i.e., sunlight). These low-energy photons break down NO_2 into NO (nitric oxide), plus a free oxygen atom. It is these free oxygen atoms, released from bondage to the nitrogen dioxide molecule, that then team up with oxygen to form ozone.

So all you need at the surface is a sunny day and enough nitrogen dioxide, and you have an instant ozone factory. Before the Industrial Revolution no one was much concerned about ozone at the surface, because it wasn't until the invention of the internal combustion engine that large amounts of nitrogen dioxide were regularly discharged into the atmosphere. But the earth didn't have to wait until the Industrial Revolution to experience air pollution. The beginnings of air pollution evolved from one of the greatest discoveries of hu-

mankind: fire. One way of looking at the problem is to
imagine a very large space. Think of something the size
of the Astrodome, big, high, spacious: the Astrodome
Earth. Now let's go back to the days of the cave people,
when humans didn't rule the earth but were merely a
small part of the evolving life forms. A couple of clans
move into the Astrodome. The place is big enough so
that the two clans may not even run into each other.
One of the clans learns about fire. It builds a fire to cook
with, and to keep warm. The other clan sees the smoke
and builds its own fire. No problem. The Astrodome is
large and spacious enough so that the smoke rises and
dissipates in the upper reaches of the dome. All goes
well for many years. More and more people, however,
are coming to the dome—the clans reproduce quickly.
Soon there are many fires throughout the dome, and
the smoke is rapidly filling up the room, displacing the
fresh air. At the surface the humans begin to notice.
They cough more, their eyes water, they don't feel as
well.

This is what is happening with the earth now. Not
just smoke, but many other gases are being released
into the atmosphere at an alarming rate. The earth is an
enclosed system, with a wonderful proclivity to cleanse
itself, but it is being taxed to the limit by the sheer
number of humans and their waste products in the
form of gases and manufactured chemicals. This is not
speculation; it is already happening. There are signs:

—In the autumn of 1988 the *New York Times*
published a story about the Jamaican palm trees
in the southeastern United States being deci-
mated by a disease known as yellowleaf fungus.

The species may disappear from America by the turn of the century. Although the cause of the disease is a known fungus, the underlying cause is the increased ozone levels in the air, which, by placing the trees under stress, pave the way for the attacking fungus.

—Forests in parts of Germany are suffering from "early autumn" syndrome: they lose their leaves by late August and early September. The cause? Increased ozone levels in the air.

—During the summer of 1988 American farmers lost between $1 billion and $2 billion in crops. The drought was a factor, but a sizable fraction of the losses from lower crop yields can be attributed to increased ozone in the atmosphere.

Modern-day air pollution began with the Industrial Revolution, when the northern hemisphere began changing dramatically. Where once stood small villages geared to service farmers, artisans, and merchants, factories sprang up—factories with their steam engines and coal-burning smokestacks, factories with a voracious appetite for fuel, and an equally prodigious output of various chemicals and gases into the air and water.

The most notable example of the effects of this revolution was the city of London during the latter years of the 19th century. Sometimes for weeks the city was hidden under a dense layer of smoke from the coal-burning factories and mills and homes. When the fog would roll up the Thames river and mix with the smoke, the mixture was called smog, and it would often be so dense that lights were needed to see even at noon.

Across the Atlantic, in the "new land" of America, similar changes were occurring. In New England, where textile factories and paper mills sprang up like mushrooms after a rainstorm, and in the Great Lakes area, where steel mills began turning out the raw product of an awakening giant, factories transformed the face of America. In the early part of the 20th century, the cities of St. Louis and Pittsburgh both experienced episodes of air pollution so severe that their residents were warned to stay indoors. These episodes of bad air were thought to be a small price for the economic miracle taking place. And indeed, the same technology that caused the problem was able to avert catastrophe and solve the most immediate threat. Soon the air over the cities began to clear, as industry built higher smokestacks, installed filters to trap particulates, and began burning "cleaner" fuels.

While the problem of "bad air" due to smoke from factories was being solved, another problem of far more complexity began to appear. This problem would prove not as easy to solve, for its causes entwined themselves within the very fabric of society.

Two cities became the axis around which the problem coalesced: Detroit and Los Angeles. Back in the first decade of the century, young Henry Ford had a vision that would revolutionize the world. Not only did he foresee a future in which every adult would own an automobile, but he envisioned a system of manufacture that would transform the nature of work. The assembly line made mass production possible, and soon the Model T Ford was chugging down the most remote dirt strip in the most far-flung parts of America.

At about the same time, on the edge of the continent, the small, quiet town of Los Angeles was beginning to stir. By the 1920s, the publicity brochures sent out by the Los Angeles Chamber of Commerce boasted of the blue sky above Los Angeles. Photos of Mt. Wilson standing snow-covered above downtown LA appeared to be a picture of paradise. Ironically, the very elements responsible for this "paradise" would soon turn the city into a modern version of Dante's inferno.

The city of Los Angeles rests in a geographical basin surrounded on three sides by mountain ranges and on the fourth by the cool Pacific Ocean. The three mountain ranges virtually isolate the basin from air currents coming from north, south, or east. The open end of the basin faces the west, the Pacific Ocean. It is these same mountains that block off the cooling ocean breezes and create the deserts that surround Los Angeles. But within the basin, within the cozy confines of the ring of mountains, the climate is almost ideal. In fact, the climate of Los Angeles is called *Mediterranean* by geographers, because of its similarity to the gentle climate in the region surrounding the Mediterranean Sea.

Such a climate is known for its mildness—no extremes of weather, sunny days and cool nights, little rainfall except during a distinct rainy season. Ideally, the way the climate of Los Angeles works is this: As the sun heats up the basin, hot air near the ground rises. As the molecules of air rise they leave a vacuum near the ground where they were. This is called a *convection current*. Since cool air is heavier than hot, the cool air over the ocean is drawn inland to replace the hot air rising. That's why in Los Angeles, unlike in similar areas at the same latitude further inland, the hottest time of day is usually around noon, with the sea breeze, caused by the convection current, cooling the temperatures by two or three in the afternoon.

Back in the 1920s this type of climate was ideal for orange groves. And indeed, the Los Angeles area back then was little more than a gathering of villages interspersed among the orange groves. Anaheim, San Bernadino, Riverside, Santa Monica, and Burbank were linked by an electrified rail system. One could ride the train from downtown Los Angeles to Santa Monica on the ocean, or the other way, to Pasadena in the foothills of the mountains. Los Angeles was a quiet oasis, a relaxed small city enjoying paradise.

Still, there was something unusual or strange about the area to be later discovered. Back in time when the first Spanish explorers had traveled around southern California, folklore had it that they heard from native Americans about a valley near the sea full of evil spirits. Native Americans had avoided the basin because when they built fires the smoke wouldn't rise as it normally did. Instead, it would hover over the fire.

But in the early part of the century, while Los Angeles remained a small city ringed by small villages, this quirk of geography went virtually unnoticed. The days were balmy, the nights cool, and seldom was a picnic or a day at the beach rained out. However, two historical events soon transpired that changed everything: the Depression and World War II. In the 1930s, while the nation and the world suffered through the worst depression in history, farmers in Texas, Oklahoma, and throughout the plains watched as their cropland turned to dust after years of drought. Meanwhile the word was out on southern California. The land of paradise beckoned: orange groves and palm trees and beautiful weather. Soon a trickle of farmers and ranchers from the drought-stricken plains began filtering through the mountain passes to the Los Angeles basin. They sent word back—"It's everything they said it was"—and soon the trickle became a stream.

By the time the Depression ended and America was preparing for war, Los Angeles had a willing and eager labor force for the aviation industry, which had just been waiting for the war to shift into high gear. More and more people began arriving in the land of sunshine. By the end of the war, Los Angeles was the fastest growing city in the country. It was a boom time.

What appeared to be a boom was actually the worst thing that could happen to the beautiful, mountain-ringed coastal basin. With a growing population the city fathers made a decision. As they saw it, the future belonged to the automobile. If the area was to support the growing population, if it was to become the city of the future, then they had to make room for the automobile, which was fast becoming a necessity for the postwar family. So they decided that the rail lines had to go. To replace them, they would build modern, wide concrete highways which would soar and swoop over the land, connecting all the villages and making automobile travel within the basin quick and easy.

But what about the Indians and the evil spirits? The smoke from the campfires that wouldn't rise? As mentioned before, the same mountains that provide the basin with its mild climate, the mountains that protect the city from the normal weather patterns, also create a situation perfect for the creation of smog. Beyond the mountains lies desert. The desert heats up while cool air flows into the basin from the Pacific. The hot air comes over the mountains and create an inversion layer, with hot air over the cool; the air that's heated at the surface has nowhere to go, and the normal convection currents are disrupted. This is what prevented the campfire smoke from rising. Given this large basin capped by an inversion layer, all you have to do is add thousands of automobiles to make an efficient ozone machine. An average day in Los Angeles starts out with fog that has rolled in from the Pacific during the nighttime hours. The sun burns the fog to a milky haze by midmorning. With the automobile exhaust building up, by midday the sunlight begins breaking down the nitrogen dioxide and freeing up the necessary oxygen atoms, which quickly attach themselves to the oxygen molecules to produce ozone. By the 1960s, the city of Los Angeles had begun issuing "ozone alerts." Schoolchildren were advised to stay indoors. The ill and elderly were advised to take caution. Driving into the city from any of the mountain passes, you couldn't fail to notice the yellow-brown haze sitting over the area, sometimes obscuring the tall buildings going up downtown. A view of Mt. Wilson from downtown then became a rare treat.

The combination of the automobile, history, and geography transformed not only Los Angeles, but the rest of the industrialized nations across the northern hemisphere in Europe and Asia as well as North America. By the end of three-quarters of a century, most cities in North America, Europe, and northern Asia had experienced episodes of ozone pollution. In the United States, during the summer of 1988, one of the hottest on record, urban smog reached record levels: 96 cities, counties, and other areas failed to meet the Environmental Protection Agency's standards, 28 more areas than in the previous year. In fact, during most of that summer, the entire United States east of the Mississippi was under an ozone cloud. Even in Bar Harbor, Maine, where the nearest industrial area is 200 miles away, rangers at Acadia National Park had to warn tourists to "take it easy" during the park's first-ever ozone alert.

Although the automobile is the primary culprit in ozone pollution, it is not the only one. And although most of the ozone pollution is caused by industrial by-products, it is not only the industrialized nations of the northern temperate zones that contribute to the problem. The earth is simply getting more crowded, and in a crowd everyone shares the blame.

In Bangkok, Thailand, if you travel outside the city to the vast, flat plain of the Chao Phraya River during the dry season, you'll come across a scene reminiscent of Cotton Mather's version of hell. There the tropical sky will have darkened as if it were night, and across the flat network of rice paddies and irrigation ditches, phalanxes of flickering flames will be marching toward the horizon. It looks like a scene depicting the end of the world, but in truth it is simply the end of the harvest season, when the farmers perform a ritual dating back to when humans first began to grow crops: burning the fields after harvest. By burning the fields, the soil is rejuvenated, ready to sow again in a few months.

Ritual crop-burning has been going on for centuries, in most of the semitropical areas adjacent to the equator. Such biomass-burning—the burning of organic material—creates ozone, but until recently it was thought that such ozone was quickly dissipated and caused no harm. Not until scientists started using satellite data did the magnitude of the ozone problem become apparent. In photos taken from satellites and the space shuttle, plumes of smoke could be seen gathering off the coast of Africa, and over South America, exactly downwind from where farmers were burning their fields. After analyzing data from satellites that measure ozone, scientists concluded that the highest ozone amounts in the tropics are located off the west coast of Africa.

Biomass burning causes ozone in the same way automobile exhaust does in the more industrialized areas. In both cases, the culprit is hydrocarbons, which are molecules composed of hydrogen and carbon and nitrogen dioxide. In the case of automobile exhaust, nitrogen dioxide and other hydrocarbon pollutants from automobile engines provide the raw material for the low-energy photons. In the case of biomass burning, it is the same gases, but in different relative amounts, that provide the raw material. Scientists now believe that the magnitude of the photochemical reaction in the tropics may be approaching the magnitude of the photochemical process in the midlatitudes of the northern hemisphere, since the rate of population growth in the tropics is two to three times that of the industrialized, developed countries. Both areas are contributing to the total background ozone levels found in the hemisphere.

In recent years ozone research and measurement have become more exact. Monitoring stations throughout the world can measure ozone levels in ambient air on a daily basis. Although measurements on a given day may look different from those on other days and from other regions, the data records are now long enough and accurate enough so that trends can be differentiated with the aid of computers.

But it wasn't always so. Ozone wasn't discovered until 1839, when the German scientist C. F. Schönberg noticed a peculiar smell while experimenting with electrically decomposing water. He wasn't necessarily looking for ozone; he was interested in the components of air. Whenever he would place an electrical charge through water, a pungent odor was released into the air. At first, Schönberg thought ozone was simply oxygen that had been negatively charged. It wasn't until years later that other scientists proved ozone to be an altered form of oxygen—an oxygen molecule with three atoms of oxygen instead of two. For many years following Schönberg's discovery, and for no apparent reason other than it was plentiful and easy to make, ozone became the darling of the scientific community. In fact, for a time during the late 19th century, ozone was actually used as a therapeutic agent for respiratory disease. Everyone wanted to measure it, study it. Monitoring stations were set up across the civilized nations of the Northern Hemisphere. But there was a problem. At the time, the instruments scientists used to measure ozone were so crude that the figures were unreliable. And most of the studies were set up on a short-term basis.

But in 1876 all that changed. At the meteorological observatory in the south of Paris, in the suburb of Montsouris, the French scientist Soret devised a means of measuring ozone content in ambient air that appeared to be consistent and reliable. By today's standards the device appears crude, the method primitive. Soret's method involved a measurable amount of air bubbled through a solution of water, arsenic potash, and potassium iodide. Ozone in the bubbling air would transform arsenite into arsenate, and the amount could be measured by the process of titration, which involves mixing the arsenate with an iodine solution and measuring the reaction.

Soret was dedicated. The measurements continued for 34 years, and to this day the "Paris studies," as they are called, represent a reference with which to compare later ozone studies. Using the Paris studies as a base, scientists feel it is safe to say that background levels of ozone in the Northern Hemisphere today are 100 to 200 percent higher than they were 100 years ago. Most of the increase has occurred since World War II, when industrialization spread and the number of automobiles worldwide increased. Since 1970, background ozone levels have been rising 1 to 2 percent per year, and the trend continues.

Ozone pollution is no longer a threat only to large urban areas. As those studies of summer campers in New Jersey and Pennsylvania show, bad air is affecting us all, whether we live in the city or the country. According to recent Environmental Protection Agency studies, the people at greatest risk are adults and children who exercise outdoors. Such damage to human health is only a small part of the ozone picture. Increased ozone levels are destroying our forests, diminishing our crops, and adding to the global warming trend. The full extent of the changes brought about by increased ozone levels at the earth's surface are not known yet.

The first step toward solving any problem is to acknowledge the problem. In the following chapters we explore the ozone problem in greater depth, and we offer possible solutions. What is obvious is that the earth has shrunk and we truly are rubbing elbows with our neighbors halfway around the world. The farmer burning his field in Zaire affects us in Maine. The steel mills in northern Indiana and Ohio affect the air over London. Realizing this is the first step. Realizing this brings us face to face with the notion of global change.

CHAPTER TWO

The Concept of Global Change

~~~~~~~~~~~~~~~~~~~~~~~~~~~~~~~~~~~~~~

The year is 2040. Sam and Melissa are both in their 60s now. They live on the coast of Maine, having moved with their families there shortly after the turn of the century. They arrived when Maine still experienced cold winters and abundant snowfall. But they consider themselves lucky now, for the state of Maine hasn't lost much of its coastline as other states along the eastern seaboard, or the Gulf Coast. Their town sits at the mouth of a tidal river, on the kind of solid granite outcroppings that give the Maine coast its distinctive look. But further south during the past 40 years the land has seen many changes. The beaches of New Jersey are only a memory. The city of New Orleans, as well as other cities along the Gulf Coast, are still livable only because of the massive systems of earthen dikes surrounding them, keeping the rising water levels out of the area. Much like the Dutch, the New Orleans residents take great pride in their ability to live and carry on merely at arm's length from disaster.

People have adapted to the changing climate. Sam and Melissa remember the harsh Maine winters with a kind of fondness now: a snow cover that would remain for three months, lakes frozen over from December to March. Now Sam plays golf in February. The summers are longer, warmer. The nippy summer nights that made Maine a haven for residents of Boston and New York are only a memory. Sam has read somewhere that Maine now has the same climate North Carolina did 100 years ago.

Out west there have been some problems. The changing climate has turned much of the fertile cropland into desert in such states as Oklahoma, Nebraska, Iowa, and Missouri. The wheat and corn belt has moved north. Canada and the Soviet Union are now the leading producers of wheat in the world. Canada's population has doubled in the past 50 years as more and more Americans have settled in the northern reaches of Manitoba and Saskatchewan, where warmer temperatures and more rain have allowed previously empty land to be placed under cultivation. Many of the residents are farmers escaping the heat and aridity of the United States' own Midwest.

Sam is lucky. The northern tier of states, where he has lived, has undergone mostly positive changes from the vast global climatic upheaval. Longer growing seasons and warmer temperatures have made Maine more self-sufficient in its food supply. The ski resorts in the mountains have abandoned the upper slopes and built year-round golf courses at the lower elevations where the rolling terrain challenges the most intrepid golfer. With the ocean temperatures warmer, many resorts have sprung up on the outer islands in the Gulf of Maine. All in all, he is happy to be living where he is, but he knows that many people are not.

The concept of global change is not something new. It has been the rallying cry for atmospheric research since the mid-1980s, when some scientists began to speculate that perhaps the climate *was* changing. Traditionally, research on global change has focused on a double-pronged attack: the observed increase in carbon dioxide (CO_2) concentrations and the depletion of the "ozone shield" in the stratosphere.

Much speculation has attended the CO_2 increase: it is this factor more than any other that is generally attributed to the "greenhouse effect," the expected warming of the global climate. The depletion of the ozone layer in the stratosphere has also been widely discussed, especially the so-called hole in the sky over Antarctica. Scientists fear that a further spread of this phenomenon could bring about worldwide epidemics of skin cancers, permanent damage to our immune system, and irreparable damage to many fragile ecosystems, especially those on the margins of life such as rain forests and tundras, because of the increase in harmful ultraviolet radiation that would occur without the protective ozone layer. But another aspect of global change has not received much attention: the danger of a global increase of ozone in the lower atmosphere, the air closest to the earth's surface, the air we breathe. So these three components—the increase of CO_2, the depletion of stratospheric ozone, and the increase of tropospheric (lower-atmosphere) ozone levels—comprise what we call the trilogy of global change. The third part of this trilogy has been more difficult to observe in the atmosphere, and the existence of such an increase in ozone levels has been at the center of scientific debate for more than a decade.

GLOBAL BUDGET OF CO_2 AND OTHER TRACE GASES

How have scientists determined that the composition of the atmosphere has indeed changed? To understand how, it is important to understand the concept of the global budget as applied to trace gases in the atmosphere. One way that may help is to think of the budget of a trace gas as analogous to a financial budget: we keep track of how much comes in and from where it originates (its sources); how much goes out (its sinks); and how much we have on hand in the atmosphere reservoir (its concentration). These three terms—*source, sinks,* and *concentration*—are the words we use to describe the movement and quantity of a trace gas.

For CO_2 the difficulty in formulating a budget is compounded by the fact that carbon dioxide is linked to the much broader concept of a biogeochemical cycle, that is, the carbon cycle. CO_2 is the dominant form of carbon in the atmosphere. On a time scale of seasons to centuries, large carbon exchanges occur between the atmosphere, the waters in the upper portion of the oceans, and the living and dead vegetation found mostly in soils. All of these can be seen as reservoirs of carbon, of roughly the same magnitude. The atmosphere loses and gains about 10 percent of its carbon content each year by exchange with the other reservoirs.

The seasonal and spatial variability of CO_2 are relatively small. Seasonal variations are largely controlled by northern hemisphere seasonal growth of plants whose rapid removal of CO_2 from the atmosphere from May to August reduces its atmospheric concentration.

Decay processes restore CO_2 to the atmosphere and occur with little variation throughout the year, though the process is accelerated during the growing season.

So from this simple picture of CO_2's biogeochemical cycle, we should expect higher concentrations in the northern hemisphere due to the greater amount of land and the greater amount of vegetation. We would also expect a small but measurable seasonal cycle of CO_2, following the seasons. Over the past several years, however, with the aid of advanced technological improvements in monitoring carbon dioxide, a different picture has begun to emerge.

The first inkling that the earth's atmosphere could be undergoing a significant change in its composition occurred in the early 1960s, on the barren top of a 9,000-foot remnant of a dormant volcano named Mauna Loa, in Hawaii. There, scientists first documented dramatic evidence of this change in the long term records of carbon dioxide measurements. These findings recorded in Hawaii offered the first conclusive indication that global change could be observed on a time scale of only a few years.

The records at Mauna Loa began in 1958 and, since then, have shown a steady rise in CO_2 concentration of about 0.2–0.3 percent per year, the figures rising from 315 ppm (parts per million) in 1958 to more than 350 ppm in the early 1990s. At first the explanations for this increase focused on the rise in use of fossil fuels. Later, scientists posited alternate explanations, one of which considered the seasonal cycle that is observed in Figure 1. As noted earlier, the seasonal cycle of CO_2 at Mauna Loa and later observed at other locations in the northern hemisphere is believed to be the result of the growth

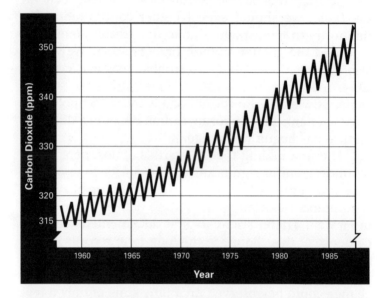

Figure 1. The carbon dioxide record from Mauna Loa since 1958. The cyclic nature of the curve is a result of the biological uptake due to photosynthesis. As plants grow more during the summer, they use more carbon dioxide, and therefore, its content in the atmosphere decreases. The increase during this time has generally been 0.2 to 0.3 percent per year. (Figure courtesy of National Oceanic and Atmospheric Administration.)

cycle of plants. In other words, scientists hypothesized that CO_2 in the atmosphere decreases during the growing season when plants absorb CO_2, and increases during the winter months because of the decay of leaves and other dead material such as bark, ground cover, and other organic materials.

Another alternative explanation, which proved consistent with the seasonal changes, has to do with the increase of deforestation, the cutting down of vast tracts of virgin forests, in the tropical regions. The tropical rain forests are being destroyed at a rapid rate, a rate sufficient to result in considerably less carbon dioxide's being consumed by photosynthesis, the process by which plants transform carbon dioxide into chlorophyll. Thus this alternative explanation is that there are fewer trees around to inhale carbon dioxide. This reduction in biomass (i.e., the burning of organic material) has become significant on a global scale, as indicated by the long-term increase observed in the Mauna Loa measurements for the atmospheric CO_2 cycle.

But how to prove it? Here is an example of how a good set of data spawned new ideas (other than the obvious). But the new ideas have still been difficult to prove. Unlike most other sciences, atmospheric studies cannot be confined to the laboratory. Scientists have found it difficult, if not impossible, to prove in the laboratory what takes place on a global scale in the atmosphere. Thus they never have found a way to prove whether deforestation in the tropics has been the primary reason for the increased CO_2 concentrations recorded at Mauna Loa.

Even today no one is sure how much deforestation contributes to the observed global increase in CO_2, not only at Mauna Loa but at other locations throughout the world between the South Pole and Alaska. On the other hand, a consensus of scientists now supports the premise that fossil fuel combustion—the burning of coal, gasoline, and kerosene in our automobiles, factories, and generating plants—is probably the major contributor to the observed global increase of CO_2.

Other processes, such as the slash-and-burn type of agriculture practiced in the tropics and the possible warming of the oceans (CO_2 is less soluble in warm water, and thus warm water results in a smaller outflow, or sink, for CO_2), may also contribute to the observed increase in carbon dioxide. Scientists can only estimate the relative contribution of these two processes, but collectively they probably contribute a small but meaningful amount to the observed increased CO_2 concentrations.

What we do know is that the global amount of CO_2 in the atmosphere has increased as first documented by the Mauna Loa records. Going back in time, we now have records obtained from ice core samples in which bubbles of air are trapped in the snow as it falls each year. As more snow falls on top of each successive layer of the compacted snow beneath it, tiny bubbles of air are literally frozen in time. The deeper below the present surface the sample is located, the older the air trapped in that compacted snow must be. Such ice core analyses confirm that CO_2 concentrations have increased dramatically since the onset of the Industrial Revolution at the end of the 19th century. Figure 2 shows that preindustrial concentrations of CO_2 were 275–285 ppm from records dating back to before 1750. Compared to the current levels of about 350 ppm, these

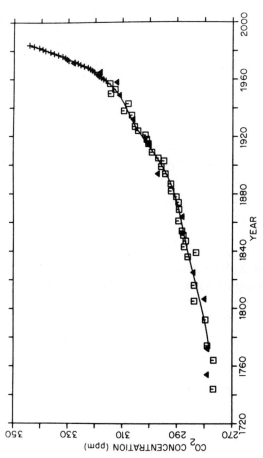

Figure 2. The increase of carbon dioxide concentration since the 18th century. The triangles and squares have been obtained from bubbles trapped in ice core samples from two independent studies using ice cores from Antarctica; the crosses are from the atmospheric measurements at Mauna Loa (Figure 1). These measurements show that the preindustrial concentration, before 1800, was about 280 ppm. (Figure from H. Oeschger and U. Siegenthaler, "How Has the Atmospheric Concentration of CO_2 Changed?" in *The Changing Atmosphere*, F. S. Rowland and I. S. A. Isaksen, eds. Chichester, England: Wiley, 1988, pp. 5–23. Reprinted with permission.)

measurements show quite dramatically that CO_2 concentrations have increased considerably during the 20th century.

But the real story isn't just the increase of CO_2 in the atmosphere. The real threat to our planet is the probability (based on theoretical calculations) that increased CO_2 will result in a warming of the earth's climate. CO_2 by itself is not a molecule that is dangerous to humankind. If all there was to worry about was the increase in carbon dioxide, life on the planet could adapt quite easily.

Other molecules in the atmosphere also have the potential of becoming greenhouse gases. All that is needed for a gas to act as a heat absorber is that it be composed of at least three atoms. Thus increases in molecules of water vapor (H_2O) or methane (CH_4) likewise will contribute to the greenhouse effect. In fact, we've probably experienced the greenhouse effect on a smaller scale when we've noticed how the nighttime air doesn't cool off as rapidly on cloudy nights as it does on clear nights. During daylight hours the ground absorbs the heat energy from the sun. After sunset, the excess heat is returned to the atmosphere. If clouds are present, some of this heat is absorbed by the water vapor molecules in the clouds. The H_2O molecules (and any molecules, for that matter) would prefer to exist in what is known as their ground state, that is, a molecular state that contains as little energy as possible.

One way to imagine this process is to visualize the three individual parts of the H_2O molecule being connected by springs that slowly stretch back and forth in the molecule's ground state. Once the molecule receives extra energy from the heat below, the springs vibrate at a faster rate. This we would call an excited state. To return to its less energetic ground state, the excited H_2O molecule discards excess heat energy in a completely random manner. An equal amount of heat is thrown off in every direction, some of it back down to earth. The downward return of some of this heat energy keeps cloudy nights warmer than clear nights.

Obviously, then, any increases in other trace gases in the atmosphere that absorb and then release heat energy could have an impact upon the earth's climate.

THE THREAT TO THE STRATOSPHERE

The second component in our global-change scenario involves the depletion of the ozone layer. The concern over the decrease of stratospheric ozone is that it will eventually result in an increase in skin cancer and may lead to other threats to human health. Satellite and ground-based measurements now confirm that the ozone layer in the stratosphere is slowly becoming thinner throughout the world, and that this effect is amplified by the existence of a "hole" in the ozone layer over Antarctica as the polar night wanes each September and October (see Figure 3).

A brief review of the stratospheric ozone problem will help place the increase in tropospheric ozone, the third component of global change, in its proper perspective. The ozone layer exists in the stratosphere (mainly between 12 and 24 miles above the earth's surface) because the sun emits radiation and because one of the major components of the atmosphere is oxygen. In fact, oxygen comprises about 21 percent of the earth's

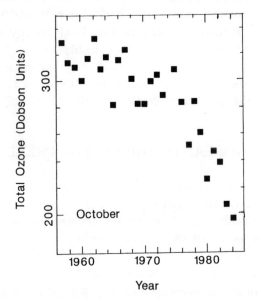

Figure 3. The average October total ozone amount in Dobson units over Halley Bay in Antarctica between 1957 and 1984. These are the data that were available when the "ozone hole" over Antarctica was first described by a team of scientists from the British Antarctic Survey who published their findings in *Nature* in 1985. One Dobson unit is 2.69×10^{16} molecules of ozone cm^{-2}. Subsequently, lower total ozone amounts were observed over Antarctica in 1985 and 1987. The 1987 October average was 120.

atmosphere, compared to 78 percent of nitrogen and 1 percent of argon, the other two primary components. In the lower atmosphere, water vapor is a major component, sometimes comprising as much as 3 percent of the atmosphere. All other gases present are in trace amounts, measured in mixing ratios such as parts per million (ppm) or parts per billion (ppb).

The radiation the sun emits and the earth's atmosphere, which receives it, can be thought of as two different continua. Although most (approximately 99 percent) of the sun's radiation is in the form of visible light and sensible heat, energy is also emitted from other parts of the electromagnetic spectrum. The electromagnetic spectrum comprises the entire scope of the sun's energy and includes the waves that bring us radio transmission and X rays, as well as visible light. This energy is "carried" by photons. The higher energy photons are primarily comprised of ultraviolet (UV) light, energetic enough not just to excite those molecules as we discussed earlier, but to break them apart completely. This process is called *photolysis.*

The earth's atmosphere can likewise be described as a continuum. Because of its gravitational pull from the earth, the atmosphere doesn't escape into outer space. Because air is a fluid, its density is considerably greater at lower altitudes. As a general rule, atmospheric density becomes less by a factor of 10 every 50,000 feet (16 kilometers) of elevation. Thus, at 50,000 feet the density of the air is about 1/10 as great as it is at the surface, at 100,000 feet about 1/100, and at 150,000 only 1/1000 as dense as it is at the surface.

Coming from outer space, a high-energy photon emitted by the sun will retain its energy until it hits something. As this photon approaches the earth, the chances for it to hit a molecule in the earth's atmosphere becomes considerably greater as the air density increases closer to the surface. If the photon hits an oxygen molecule, which is comprised of two oxygen atoms, it will photolyze this molecule into its two atoms. The process can be described by the chemical equation:

$$O_2 + \text{(high energy) photon} \rightarrow O + O$$

Once the oxygen atoms are "free," they can recombine with an oxygen molecule to form ozone (O_3). But if they are to do this, a third neutral molecule must show up to absorb the excess energy the oxygen atoms have recently acquired after being "zapped" by the high-energy photon. In most cases this third molecule is nitrogen (N_2), but it can be either another oxygen molecule or an argon (Ar) atom. Chemically, the process is described:

$$O + O_2 (+M) \rightarrow O_3 (+M)$$

where M is either N_2, O_2, or Ar.

So at the very high levels of the atmosphere, where there is relatively little N_2, O_2, or Ar present, there is much more atomic oxygen (O) than ozone (O_3). As we get lower in the atmosphere, O_3 becomes more abundant, relative to O. Again, however, we should note that atomic oxygen and ozone always remain trace constituents of the atmosphere, never reaching concentrations above the part per million range. Eventually, all the high-energy photons will collide with an oxygen molecule, and by the time the light penetrates down to

12 to 15 miles, no more of the high-energy photons can get through the earth's atmosphere, since they've all been absorbed by the abundant oxygen molecules above.

Once the ozone is formed, it can likewise be photolyzed by UV radiation. The process might look like this:

$$O_3 + photon \rightarrow O + O_2$$

Because O_3 is a less stable molecule than O_2, the amount of energy required to break apart the ozone molecule is considerably less than the amount of energy required to break apart the oxygen molecule. This instability allows ozone to absorb radiation in the ultraviolet part of the electromagnetic spectrum directly adjacent to the visible part of the spectrum. It is this UV radiation that gives us our suntans (and sunburns) in the summer and that is blocked out exclusively by ozone molecules. The more ozone present between us and the sun, the less radiation from this end of the spectrum reaches the ground.

In the stratosphere, a reaction can also take place whereby atomic oxygen, the oxygen molecule composed of only one atom, reacts with ozone to give back the normal type of molecular oxygen that is so necessary for life:

$$O + O_3 \rightarrow O_2 + O_2$$

This interplay among the various forms of oxygen in the stratosphere was first described by Sydney Chapman (Figure 4), a British physicist, in 1930. This set of reac-

Figure 4. Sydney Chapman (1888–1970), the British physicist who first theorized the existence of the ozone layer in a 1930 scientific paper. His photochemical theory of stratospheric ozone formation was described by the use of four reactions involving various forms of pure oxygen. (Photograph courtesy of the National Center for Atmospheric Research of the National Science Foundation.)

tions, which is commonly referred to as *Chapman chemistry*, describes in relatively simple terms why the ozone is located where it is, namely, in a region where the air is very thin so that high-energy photons, which are needed to break apart oxygen molecules, are still plentiful and, at the same time, the air is dense enough so these oxygen atoms can combine with oxygen molecules to make ozone.

At higher altitudes, the air is not dense enough for the oxygen atoms to react simultaneously with an oxygen molecule *and* a third molecule. At lower altitudes, none of the high-energy photons that are necessary to break apart the oxygen molecules are still around. They have already collided with an oxygen molecule above.

Through the 1950s, scientists proposed little modification of Chapman's chemical schemes to explain the chemistry in the stratosphere. During this time, important progress in atmospheric chemistry took place in the laboratory in the specialized field of chemical kinetics, which is most concerned with quantifying the rate at which chemical reactions take place in the atmosphere. Thus, even though Chapman proposed that ozone reacts with atomic oxygen to give back two diatomic (i.e., normal) oxygen molecules, it was important to discover how fast that reaction took place in the atmosphere in order to know with a higher degree of certainty how much ozone was present in the atmosphere. During the three decades following his classic 1930 paper, Chapman's hypothesis was still found to be correct, but as the science grew to a more sophisticated quantitative state, the application of laboratory chemical kinetics began to indicate that Chapman chemistry did not completely explain the amount of ozone in the stratosphere, now being measured on a more routine basis.

As the field of chemical kinetics advanced, not only did some of the basic reactions become quantified for the first time, but many other chemical reactions that hadn't been considered before were found to take place fast enough in the atmosphere to have an influence on the amount of ozone calculated to be present. In particular, the 1950s saw enormous strides taken in understanding the importance of the hydroxyl radical (OH) in atmospheric chemistry. Radicals are similar to molecules: they both are composed of several atoms which are bonded to each other. Radicals differ from molecules in their number of electrons and protons so, unlike molecules, radicals have an electric charge, either positive or negative. In the case of the hydroxyl (*hydro*gen + *oxy*gen) radical, there are 10 electrons and only 9 protons (8 from oxygen and 1 from hydrogen); thus it has a negative charge. It's this electrical charge that makes the radical so reactive.

A modification of Chapman's chemistry was proposed in a 1964 paper in which chemical reactions incorporating the hydroxyl radical were shown to be important in determining the amount of ozone in the stratosphere. The inclusion of these reactions reduced the amount of ozone in theory and thus resulted in better agreement between theoretical calculations and the observed amount of ozone in the stratosphere. This particular set of chemical reactions involving both oxygen and hydrogen components became known as *wet photochemistry* because water vapor had to be present to serve as the initial reservoir necessary for the formation of OH. Two hydroxyl radicals were formed when energetic oxygen atoms reacted with a water vapor molecule:

$$O + H_2O \rightarrow OH + OH$$

The presence of the hydroxyl radical greatly complicated the chemistry in the stratosphere by introducing reactions that can indirectly and directly result in both the production and the destruction of ozone in the stratosphere. For example, in the above reaction the amount of ozone produced in the stratosphere is slowed down because atomic oxygen (O) can react with H_2O rather than with O_2 and a neutral molecule to make O_3. Conversely, the amount of ozone destroyed through its reaction with atomic oxygen:

$$O + O_3 \rightarrow O_2 + O_2$$

is likewise slowed. The net effect of all this is complicated, but basically it means that the initial reaction of atomic oxygen with water vapor can result in either more or less ozone in the stratosphere.

Once OH is formed, it can react with ozone:

$$OH + O_3 \rightarrow HO_2 + O_2$$

which likewise results in a destruction of ozone. In addition, the generation of another radical (HO_2, called the *hydroperoxy radical*) can also react with ozone to regenerate OH radicals:

$$HO_2 + O_3 \rightarrow OH + 2O_2$$

With the measurement of these reaction rates in the laboratory, the picture of stratospheric chemistry became much more complicated than had been previously explained by Chapman's pure oxygen chemistry. Furthermore, the inclusion of these radicals in a set of calculations that were being used to describe the distribution of ozone in the stratosphere yielded a computed distribution that was more consistent with the growing data base of ozone measurements. However, it is easy to see how the science of atmospheric chemistry became considerably more complicated once additional molecules had to be considered to explain the distribution of ozone in the stratosphere.

A further refinement of stratospheric chemistry came about in 1970 when Paul Crutzen, then of the University of Stockholm, and Harold Johnston, of the University of California at Berkeley, independently introduced theories proposing that molecules that contain nitrogen, analogous to those that contain hydrogen, play an important role in the chemistry of the stratosphere. Their proposals supported a growing concern at the time that direct injection of nitrogen oxides by high-flying aircraft might seriously impact the chemistry of the ozone layer.

Prior to Johnston's and Crutzen's studies, the ozone layer was primarily a topic of only academic interest. Then the atmosphere became a subject of interest to more than the scientific community when a consortium of British and French companies, as well as some U.S. aircraft manufacturers, considered building a fleet of supersonic aircraft (SSTs) to fly at altitudes that would take the planes into the lower reaches of the stratosphere. As a result of this concern, the U.S. Department of Transportation formed the Climatic Impact Assessment Program (CIAP). What CIAP was concerned about was the effect of a fleet of SSTs emitting large quantities of water vapor and nitrogen oxides into the lower stratosphere. The potential threat to the ozone layer was thought important enough for the United States to scrap its own plans to develop a fleet of such aircraft. The British and French went ahead, but the size of the fleet is still so small that in the 1990s the impact of such planes flying through the stratosphere is still negligible.

In 1974 a pair of young scientists from the University of Michigan, Richard Stolarski and Ralph Cicerone, wrote a paper in which they suggested that chlorine atoms in the stratosphere likewise might play an important role in ozone chemistry. They had been working on a study funded by NASA to study the potential effects of space shuttle exhaust on the ozone layer. One of the components of this exhaust is hydrochloric acid (HCl), which can be easily photolyzed by the ultraviolet radiation in the stratosphere, releasing free chlorine atoms (Cl). Stolarski and Cicerone identified the following reactions in the stratosphere which would destroy ozone:

$$Cl + O_3 \rightarrow ClO + O_2$$

followed by

$$ClO + O \rightarrow Cl + O_2$$

The result of the sum of these two reactions is:

$$O_3 + O \rightarrow O_2 + O_2$$

More importantly, the chlorine atom that first destroyed the ozone molecule is returned to the atmosphere to destroy more ozone. Such a sequence of reactions is called a *catalytic cycle of ozone destruction*, where the chlorine serves as a catalyst to destroy the ozone. Theoretical calculations based on laboratory measurements of the rates of these various atmospheric reactions indicate that each chlorine atom goes through this cycle 10,000 to 100,000 times before the Cl atom reacts with something other than an ozone molecule in the upper atmosphere.

The amount of ozone destroyed by space shuttle exhaust in the stratosphere turns out to be relatively insignificant as long as the space shuttle and other space rockets are launched as infrequently as they have been since the onset of the space program in the 1960s. That was the good news. The bad news was that there is a source of manufactured chlorine that creates an important perturbation in the natural chemistry of the stratosphere.

In 1974, Mario Molina and F. Sherwood Rowland, of the University of California at Irvine, published a paper in the journal *Nature,* which showed that chlorofluorocarbon (CFC) molecules can serve as an important source of free chlorine atoms in the stratosphere. These manufactured molecules are extremely stable in the lower atmosphere and can only be broken down by the intense UV radiation found in the upper stratosphere.

There are two primary forms of CFCs. CFC-12 (chemical formula CF_2Cl_2), which is used primarily as a refrigerant in air-conditioning systems, is comprised of one carbon atom surrounded by two atoms of chlorine and two atoms of fluorine. CFC-11 (chemical formula $CFCl_3$), which was being used in the 1960s and 1970s as a propellant for aerosol spray cans, has a carbon atom surrounded by three chlorine atoms and one fluorine atom. Upon reaching sufficient altitudes, they can photolyze to release chlorine:

$$CFCl_3 + \text{(high-energy) photon} \rightarrow 3Cl + \text{products}$$

Analogously, CFC-12 breaks down in the upper stratosphere to release its chlorine atoms.

Without these manufactured chlorine molecules, there would be almost no other chlorine present in the stratosphere. The trend in the amount of these two chemicals that have been measured in the lower atmosphere is shown in Figure 5.

As can be seen from this figure, the amount of CFC-12 has increased from less than 200 parts per trillion (pptv) in 1975 to more than 400 pptv in 1987. When the chlorofluorocarbons were first measured in the atmosphere in the 1960s, their concentrations were less than 100 pptv, by volume. In the early parts of this century, their concentrations were zero. Although there have been attempts recently to ban the use of these chlorofluorocarbons, such as those used in aerosol cans, whether such a ban will help is difficult to predict, since CFCs are not broken down very quickly. As measurement techniques were developed in the 1970s, scientists discovered that the amount of chlorine in the stratosphere has increased by a factor of 2 to 3 since the 1970s as CFCs slowly drifted up to the stratosphere and released their chlorine.

In 1986 a panel of 130 scientists from throughout

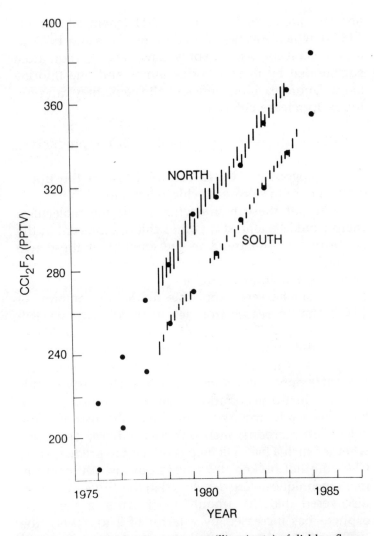

Figure 5. Concentration in parts per trillion (pptv) of dichlorofluoro-methane (CCl_2F_2) between 1978 and 1985 is shown for measurements in Ireland (vertical lines labeled "North") and Tasmania (ver-

the world organized to determine and quantify the extent to which the ozone in the stratosphere had been depleted. Furthermore, if a decrease of ozone in the stratosphere could be definitively established, this panel was charged with the responsibility of finding out whether the decrease could be attributed to natural or anthropogenic (human-produced) causes. Prior to the establishment of this fact-finding panel, several scientific papers had been published which reported that ozone amounts in the stratosphere had been decreasing in the late 1970s and early 1980s. There were, however, several possible explanations for the observed decreasing trend.

One distinct possibility was that one of the satellite instruments that had been making continuous measurements between 1978 and 1987 had been drifting in its calibration. This instrument, the total-ozone-mapping spectrometer, commonly referred to as TOMS, had been providing scientists with daily maps of the global distribution of ozone in the entire atmosphere. Although it had been designed to provide data for only two years, the TOMS had already been operating flawlessly for more than seven years when the panel decided to examine whether or not the absolute numbers it had been sending back to earth were systematically lower because of some kind of electronic drift of

---

tical lines labeled "South") together with January flask data (grab samples) from Oregon (upper dots) and South Pole (lower dots). (Figure from F. S. Rowland, "The Role of Halocarbons in Stratospheric Ozone Depletion," in *Ozone Depletion, Greenhouse Gases, and Climate Changes.* Washington: National Academy Press, 1989, pp. 121–140. Reprinted with permission.)

its sensor. Thus the panel was given the responsibility of reexamining the TOMS data and comparing these data with other total ozone measurements from ground-based instrumentation. *Total ozone* is defined as the total amount of ozone between the earth's surface and the top of the atmosphere.

Another possibility that would explain the observed decrease of total ozone during this period was a phenomenon called *solar variability*. The sun undergoes an 11-year cycle and the number of sunspots observed on the sun vary on a regular 11-year cycle. Actually, sunspots are storms on the surface of the sun. A manifestation of these storms is the emission of excess radiation during periods of high activity. It's well known that during periods of strong sunspot activity the excess electromagnetic radiation emitted from the sunspots manages to intercept the earth's magnetic field, causing a disruption of radio communication on earth and sometimes a display of auroras at latitudes where they are not normally seen. As these energetic particles enter the earth's atmosphere, they react with nitrogen and oxygen molecules in the upper atmosphere, and the result is the formation of excess nitric oxide (NO). In principle, the presence of this excess nitric oxide in the upper stratosphere could result in a depletion of ozone. Thus, since the frequency and magnitude of these events had been tied to the 11-year solar cycle, it was thought possible that the decrease of ozone observed in the late 1970s and early 1980s was part of a natural variability of ozone as a result of the solar cycle. The ozone panel was charged with the responsibility of quantifying how much of an effect, if any, solar activity had had on the observed ozone trend during this time.

Finally, there was another event of natural origin that could have had an impact on the ozone trend during this period. In 1982, the El Chichón volcano erupted. The eruption was the largest of the decade, releasing enormous amounts of volcanic debris into the stratosphere. The magnitude of such an eruption can be better understood when comparing the El Chichón event with the eruption of Mt. St. Helens in 1980 (see Figure 6). The El Chichón eruption put more than 12 megatons of dust and gases into the stratosphere whereas Mt. St. Helens put only about one-half megaton of matter into the stratosphere. Once dust from such a volcano reaches the stratosphere, it remains there for years. Scientists know that the presence of volcanic matter in the stratosphere causes incoming light to be reflected differently from times when volcanic matter is not present in the stratosphere. Possibly, this dust layer in the stratosphere had resulted in erroneous measurements, from both space-borne and ground-based instruments. It's also possible that some of the chemicals injected into the stratosphere by the El Chichón eruption caused a depletion of the ozone in the stratosphere.

In 1988 the panel released its findings. After sorting through all the data, they concluded that the observed decrease between 1978 and 1987 was indeed real, although they did establish that the TOMS instrument had also been undergoing a drift in its calibration since 1983. They concluded that ozone had decreased by about 3 percent between 1978 and 1987 in the southern hemisphere, and by about 2 percent in the northern hemisphere. This decrease was about half what the TOMS had measured. The panel also concluded that the natural

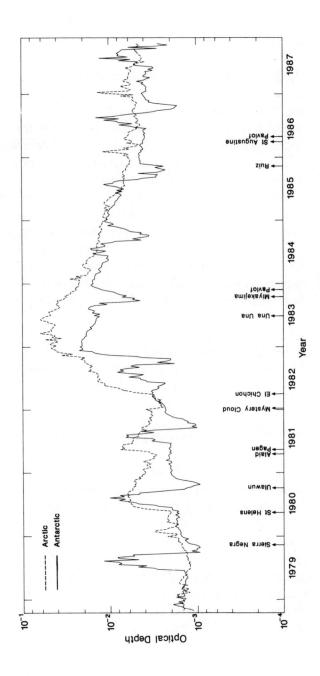

Figure 6. The optical depth of the stratosphere determined from the SAM II satellite measurements between 1978 and 1987. The optical depth is a measure of the amount of particulate matter (primarily dust) in the atmosphere. The dashed line shows the measurements over the Arctic, whereas the solid line is over the Antarctic. Particularly noteworthy is the large increase in the optical depth in 1982 (especially in the northern hemisphere) after the eruption of El Chichón in Mexico. A smaller jump is seen in 1980 following the eruption of Mt. St. Helens in Oregon, which only showed an increase in the optical depth over the Arctic. The quasi-regular peaks seen in the Antarctic are primarily the result of polar stratospheric clouds, whose formation occurs primarily in late winter and early spring in the southern hemisphere. At one time, there was great concern that the dirtiness of the stratosphere following the El Chichón eruption had led to erroneous measurements of stratospheric ozone by both ground-based and satellite instruments. (Figure courtesy of M. Patrick McCormick, NASA Langley Research Center.)

perturbations due to either the solar cycle or the eruption of El Chichón were not the cause of the observed decline. They concluded that the increased chlorine in the stratosphere over the past decade had been the primary reason for the depletion of the ozone layer.

While all the theories showed that there should be a decline of ozone due to the input of these chlorinated compounds, none of them suggested that there should be a region particularly sensitive to the impact of increased chlorine. In 1985, scientists from the British Antarctic Survey reported in the journal *Nature* their discovery of a "hole" in the ozone layer over Antarctica as the polar night ended in September and October. The scientists, led by Joe Farman, described a series of measurements using ozonesondes (a balloon-borne ozone instrument that telemeters ozone concentrations back to the ground via a radio transmitter and receiver), which showed that ozone levels were barely half of their normal levels over Halley Bay during this particular time of year.

Figures 7 and 8 show an example of this dramatic effect. *More than 90 percent of the ozone between the altitudes of 15 and 20 kilometers had disappeared over a six-week period.* Even the chlorine chemistry that Stolarski and Cicerone had described, and that Molina and Rowland had warned about in 1974, could not have been responsible for such a dramatic decrease over such a short period of time. So what was responsible?

As it turned out, the mechanism responsible for such a dramatic depletion in ozone as was found in the Antarctic ozone "hole" involves a unique kind of chemistry that wasn't even known to exist until detailed observations of the structure, extent, and chemistry of the Antarctic region were obtained. These observations

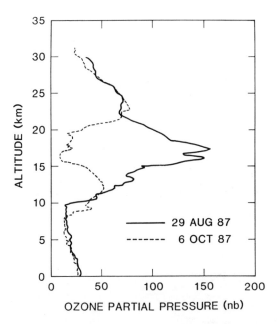

Figure 7. Two ozone profiles obtained from ozonesonde launched at McMurdo station in Antarctica. More than 90 percent of the ozone located between 15 and 20 kilometers, the center of the ozone layer, disappeared during a six-week period in 1987, the most dramatic decrease ever observed up to that time. The low stratospheric ozone concentrations (plotted here in nanobars, nb) persisted through the month of October. A nanobar is a unit of atmospheric pressure (10^{-9} bar). The pressure at the earth's surface is 1 bar. A more common unit of measuring pressure is the millibar (mb), which is 10^{-3} bars. (Figure from R. T. Watson, "Stratospheric Ozone Depletion: Antarctic Process," in *Ozone Depletion, Greenhouse Gases, and Climate Change*. Washington: National Academy of Science Press, 1989, pp. 19–32. Reprinted with permission.)

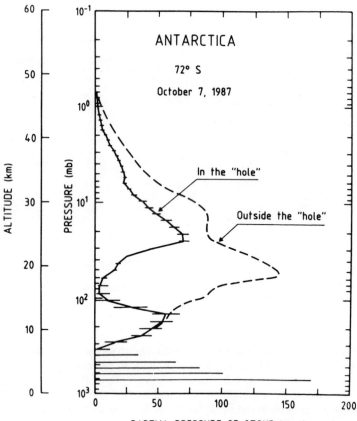

Figure 8. The sharp definition of the ozone hole. These two vertical profiles of ozone were obtained within a few hundred miles of each other on October 7, 1987, over Antarctica. The vertical coordinate is shown in both altitude (km) and pressure (millibars, mb). The concentration is plotted in nanobars (nbar). (Figure from "Group Report, Changes in Antarctic Ozone," in *The Changing Atmosphere*, F. S. Rowland and I. S. A. Isaksen, eds. Chichester, England: Wiley, 1988, pp. 235–258. Reprinted with permission.)

weren't even available until a comprehensive expedition involving two aircraft, ground-based instruments, and special balloon observations was undertaken in August through October of 1987.

What the expedition discovered was that the chemical processes responsible for the ozone hole rely in part upon the presence of a phenomenon found in the polar stratosphere called *polar stratospheric clouds* (PSCs). PSCs form in the coldest regions of the atmosphere at temperatures below −80°C (−112°F). Such temperatures occur most frequently in the stratosphere over the South Pole, where the circulation of the atmosphere isolates this air from mixing with air from the lower latitudes which would warm it. PSCs also form in the Northern Hemisphere and were identified as far back as the late 1800s, when they were called "mother-of-pearl" or nacreous clouds by Norwegian scientists. Studies in the early 1920s and the 1930s showed that such clouds form high in the stratosphere when certain atmospheric circulation patterns are present. But there is a difference between the PSCs we see now and the ones seen by the Norwegians. Whereas the mother-of-pearl clouds had a negligible impact on the chemistry of the stratosphere when they were first observed, the presence of the clouds today results in a perturbation of the chemistry in the stratosphere because of the high levels of chlorine in the atmosphere that were not present even a decade ago. As a result of this large amount of unnatural chlorine in the stratosphere, PSCs now provide efficient reaction surfaces that liberate reactive chlorine atoms that otherwise would have been tied up in benign molecules such as chlorine nitrate or hydrochloric acid. In addition, some of the particles that comprise the PSCs eventually grow so large that they settle to lower altitudes in the stratosphere or even into the troposphere.

When this process occurs, most of the nitrogen in the stratosphere is removed locally, allowing the catalytic cycle of ozone destruction by the chlorine left behind to proceed much more efficiently. Thus, an entirely new type of chemistry takes place on these very cold particles that make up the PSCs—so new it had never been simulated in the laboratory. However, one of the by-products of this ice-particle chemistry is that an extremely large amount of chlorine is liberated and then released to the atmosphere. As we have seen, chlorine has an appetite for the ozone molecule, which results in a very efficient process for removing ozone in the stratosphere near the PSCs.

But again, as with the first part of the trilogy, the problem with the second component of the global change trilogy isn't the decrease of ozone in the stratosphere *per se*. The fear is that the smaller amount of ozone in the stratosphere will allow more ultraviolet radiation to reach the surface. In turn, this radiation will cause environmental and human health damage. The stratospheric ozone threat may be even more serious than was first described by the initial doom-and-gloom theories put forth in the early 1970s. Back then scientists weren't aware of the chemistry that was accelerated by the existence of the clouds high in the stratosphere.

Similar clouds, when first observed by the Scandinavians, were beheld as one of the truly beautiful phenomena unique to polar regions. When they were first observed, they truly were magnificent. However, back then there was virtually no chlorine in the stratosphere. The addition of chlorine from manufactured chemicals in the last half of the 20th century has made these stunning phenomena a lethal reminder of humankind's contribution to the ozone layer.

CHAPTER THREE

The Trilogy of Global Change

CHAPTER THREE

The Biology of Global Change

In 1974 a small jet aircraft took off from Mainz, West Germany. Aboard it was a young scientist named Wolfgang Seiler (see Figure 9) with a relatively new instrument, capable of measuring amounts of carbon monoxide (CO). The instrument was revolutionary for two reasons: First, it could measure relatively low amounts of CO, and second, it could give readings in a matter of seconds. This second factor was important, since it meant that a scientist flying through the atmosphere would have available instantaneous readings of the air outside the aircraft. As part of his doctoral research in the late 1960s, Seiler had measured carbon monoxide levels throughout the world and had become the undisputed expert on this particular trace gas. Now, as a research scientist at the Max Planck Institute for Chemistry in Mainz, Seiler had just received funding from the German government to embark on a six-week series of aircraft flights to obtain a clearer picture of the distribution of carbon monoxide throughout the troposphere and the lower stratosphere. He was unaware that his measurements would provide the scientific community

Figure 9. Wolfgang Seiler (1940–) (center), an atmospheric chemist who provided the first comprehensive global survey of the distribution of carbon monoxide. Seiler developed a new instrument to measure carbon monoxide as part of his doctoral research at Johannes Gutenberg University in Mainz, West Germany. He continued his work in the 1970s and 1980s as a research scientist at the Max Planck Institute for Chemistry in Mainz. In 1986, he became director of the Fraunhofer Institute for Atmospheric Environmental Research in Garmisch-Partenkirchen, West Germany.

with important new insights into the origin of ozone in the troposphere.

Before Seiler's new instrument, the only way to measure carbon monoxide in a nonurban, unpolluted atmosphere, was by means of a process that required at least 10 to 20 minutes. Seiler had already mapped the CO concentrations at ground level from a cruise ship as part of his dissertation research, and in 1974 he was to begin a six-week journey aboard a specially equipped jet plane. His purpose? To obtain a better understanding of how the amount of carbon monoxide in the atmosphere varies from one location to another and from one altitude to another. He was particularly interested in how much of it is in the atmosphere, and if there is more in the northern hemisphere stratosphere than in the southern hemisphere stratosphere. Seiler's journey would help answer such questions.

The airplane he was aboard could fly at altitudes above 40,000 feet, high enough to get him into the stratosphere at mid and high latitudes. To enable him to determine when he was entering the stratosphere, he had to measure ozone levels. High ozone concentrations at altitudes of 40,000 feet would tell him at a glance that he had reached the stratosphere, and any sudden increase in ozone levels as he ascended would identify the boundary between the troposphere and the stratosphere. Since the new, reliable ozone-measuring devices were readily available, and the response time was only a matter of seconds, Seiler was all set. As he flew off for his around-the-hemisphere journey, all he had to do was watch the needles of his two instruments—one the ozone-measuring device, and other his device for measuring carbon monoxide.

According to his traditional scientific education, the two needles should have moved dramatically in opposite directions as he entered the stratosphere—the ozone levels rising while the carbon monoxide levels dropped. In the remote troposphere, far away from urban centers, the needles on the two instruments should also have moved in opposite directions, since any high value of ozone in these pristine regions of the troposphere should have originated in the stratosphere. Such stratospheric air, even though considerably diluted, should also have contained lower concentrations of carbon monoxide.

But something happened during those six weeks that didn't fit into Seiler's expectations. At certain times while ascending or descending through the troposphere, Seiler observed his needles moving in the same direction. He responded by sometimes scribbling a note in German on his chart paper, sometimes recording such occurrences in his logbook. But when he returned to Mainz in August of 1974, he found himself overcommitted to new projects, without the time to complete detailed analyses of his six weeks of data. It wasn't until five years later that the mystery of the needles swinging in the same direction was examined in detail, and then only because of a chance meeting thousands of miles from Germany.

MAKING OZONE ON THE WAY TO CO_2

The evidence now is fairly convincing that tropospheric ozone—ozone levels in the air we breathe—is increasing everywhere in the northern hemisphere. The troposphere is generally defined as the part of the atmosphere where weather occurs, typically comprising the lowest six to nine miles at midlatitudes. Our main premise here is that the consequences of such an increase are more far-reaching than increases in any of the other trace gases, such as carbon dioxide and methane, that ideas about global change have focused on to date.

The story behind tropospheric ozone research is considerably more complex than that behind the increase of CO_2 or the depletion of the ozone layer in the stratosphere. Unlike those stories, the tropospheric ozone story had to battle against an entrenched understanding of atmospheric chemistry that had prevailed for nearly four decades.

In recent years the newspapers and magazines have been full of stories about the increase in atmospheric carbon dioxide (CO_2) due to the burning of fossil fuel. The main culprit seems to be the combustion processes using coal, petroleum, or natural gas. Petroleum and natural gas are hydrocarbons; that is, they are comprised of carbon and hydrogen atoms. Natural gas is mostly methane, the simplest of the alkane family of hydrocarbons: its molecular structure has one carbon atom at the center surrounded by four hydrogen atoms. In an alkane, each carbon has four places to which hydrogen or other carbon atoms can attach. The next alkane is ethane; it has two carbon atoms and six hydrogen atoms. The two molecules look like this:

```
        H                H   H
        |                |   |
   H – C – H        H – C – C – H
        |                |   |
        H                H   H

     Methane            Ethane
```

Both of these molecules exist as gases under typical atmospheric temperatures and pressure. The next two alkanes, propane and butane, consist of three and four carbon atoms, respectively. Under normal atmospheric pressure, they exist as gases, but they can easily be transformed into a liquid state if they are kept under enough pressure.

The next several alkanes consist of five through nine carbon atoms and are the components we know as gasoline. They are pentane, hexane, heptane, octane, and nonane, in the order of increasing numbers of carbon atoms. The most common component of gasoline is octane, a molecule comprised of 8 carbon atoms and 18 hydrogen atoms. As the carbon chains in the alkane family become longer, they make up such fuels as diesel and heating oil.

Once these alkane molecules become longer, the carbon and hydrogen atoms they are composed of can arrange themselves in several different configurations. For example, an octane molecule can look like this:

$$
\begin{array}{cccccccc}
H & H & H & H & H & H & H & H \\
| & | & | & | & | & | & | & | \\
H-C & -\,C & -\,C & -\,C & -\,C & -\,C & -\,C & -\,C-H \\
| & | & | & | & | & | & | & | \\
H & H & H & H & H & H & H & H \\
\end{array}
$$

which is called *normal octane*. An octane molecule can also have its carbon and hydrogen atoms arranged to look like the molecule depicted on the next page, which is called an *iso-octane*. Such molecules, even though they have the same number of atoms as the normal straight-chained octane, are called *isomers* of the normal molecule. *Isomer* is the name given to any of two or more

```
              H                    H
              |                    |
      HH – C – H   HH – C – H   H
        |   |   |      |   |   |      |
  H – C   –   C   –   C   —   C   —   C – H
        |   |   |      |   |   |      |
      HH – C – H   H      H      H
              |
              H
```

chemical compounds composed of the same constituents but differing in their structure. In one sense the two molecules are the same (both contain 8 carbon and 18 hydrogen atoms), but because of their differing structures, they possess slightly different properties. The main difference is that the iso-octane ignites at a slightly higher temperature than the normal octane molecule.

This is a favorable property in your automobile engine, causing it to perform better. When the science of combustion engineering became more sophisticated in the 1930s and 1940s, the automotive engineers studying the nitty-gritty of how fuel burns in an auto engine devised an arbitrary scale to measure the relative properties of various kinds of fuels. They determined that an automobile engine performed very well on a mixture that had the properties of pure iso-octane. Such a mixture was given an "octane rating" of 100.

If the gasoline contained molecules that would ignite prematurely in the combustion chamber of the engine, the octane rating of the gasoline was lower. As a reference point on the lower end of the scale, the engineers determined that a mixture having the combustion properties of normal heptane, a straight-chained hydrocarbon with 7 carbon atoms and 16 hydrogen atoms, would be given a rating of 0.

The gasoline we get now from the corner gas pump has been refined and processed to burn efficiently with an octane rating that reflects how well it burns in an automobile engine. To achieve such an octane rating, certain other chemical additives are usually "blended" into the gasoline. One common additive used in the past was lead tetraethyl. This molecule looked very much like an octane molecule (it had eight carbon atoms), but when added to gasoline in small amounts, it increased the octane rating by causing combustion to occur slightly slower in the engine. The "knocking" that we often experience in our cars results when the fuel ignites before it should in the combustion chamber. Higher-octane-rated gasoline requires higher temperatures to start combustion and requires an infinitesimally longer amount of time to burn than gasoline with lower octane ratings. Eliminating these undesirable properties in the gasoline results in an engine running without the "knock."

Eventually tetraethyl was banned from use in gasoline by the U.S. Environmental Protection Agency because of the toxic effects of the airborne lead that came out in the exhaust. In general, even without using lead additives, we are paying more for higher octane gasoline today because more of these combustion-enhancing compounds have been added.

All of these hydrocarbons liberate heat when they combine with oxygen. We call this oxidation process *combustion*, and we use the energy released by this process to power our automobiles, heat our homes, or cook our food. Eventually, each carbon atom becomes carbon dioxide (CO_2) and each hydrogen atom becomes water (H_2O). In terms of octane, it would take 25 oxygen atoms to oxidize two octane molecules completely: in general, the completeness of the combustion process is limited by the amount of oxygen available to react with the hydrocarbon. The purpose of a carburetion system is to mix the fuel and oxygen as completely as possible before the combustion takes place in the combustion chamber (piston) of a gasoline or any other internal-combustion engine. The goal of the combustion process is to convert every carbon atom into carbon dioxide and every hydrogen atom to water vapor.

By now it must be obvious that it takes a great deal of oxygen to completely oxidize the hydrocarbons powering an automobile engine. In reality, complete oxidation does not occur, and the exhaust from the engine contains gases other than CO_2 and H_2O. Some of these exhaust products are hydrocarbons that have not been burned, or oxidized, at all. Other exhaust products include hydrocarbons that have undergone incomplete combustion. Some of these gases can be partially oxidized. Other exhaust products can be shorter chained hydrocarbons which have been formed by the longer chained hydrocarbons' breaking apart, a process called pyrolysis. The most prevelant partially oxidized gas is carbon monoxide (CO). Pentane (C_5H_{12}) is one of the most abundant hydrocarbons found in the urban atmosphere. Once it is released to the atmosphere, pentane is most likely to react to with our old friend the hydroxyl (OH) radical and begin its atmospheric oxidation process through the reaction:

$$C_5H_{12} + OH \rightarrow C_5H_{11} + H_2O$$

One product is a water molecule; the other product looks like this:

$$
\begin{array}{ccccc}
\text{H} & \text{H} & \text{H} & \text{H} & \text{H} \\
| & | & | & | & | \\
\text{H}-\text{C} - \text{C} - \text{C} - \text{C} - \text{C} - \\
| & | & | & | & | \\
\text{H} & \text{H} & \text{H} & \text{H} & \text{H}
\end{array}
$$

Notice the end of the molecule without the hydrogen atom. That's like an open wound, where the hydrogen has been stripped from the end carbon. This fragment is highly reactive at the spot where the hydrogen atom has been stripped away, and it is also a radical. Most radicals of this type will continue the oxidation process by seeking an oxygen molecule to attack at the end. After such a process, the radical looks like this:

$$
\begin{array}{ccccc}
\text{H} & \text{H} & \text{H} & \text{H} & \text{H} \\
| & | & | & | & | \\
\text{H}-\text{C} - \text{C} - \text{C} - \text{C} - \text{C} - \text{O} - \text{O} \\
| & | & | & | & | \\
\text{H} & \text{H} & \text{H} & \text{H} & \text{H}
\end{array}
$$

Any radical that has an oxygen molecule attached to it in the above manner is called a *peroxy radical*. For simplicity, these fragments are labeled RO_2 radicals, where the letter R stands for any hydrocarbon fragment missing its end hydrogen atom. It can have 1 carbon and 3 hydrogen atoms, 2 carbons and 5 hydrogen atoms, or in the above example, 5 carbon and 11 hydrogen atoms. Each carbon atom in these compounds will eventually become carbon dioxide: the original goal of the combustion process. However, this oxidation process now continues to take place in the atmosphere, and not just within the combustion chamber. These peroxy radicals originating from the incomplete combustion of the hydrocarbons are the necessary precursors of the efficient production of photochemical smog and ozone.

Another by-product of combustion is nitric oxide (NO). This trace gas is formed as a result of the high temperatures that exist in the combustion chamber of an internal combustion engine. The extreme heat found there is capable of breaking apart the stable nitrogen and oxygen molecules N_2 and O_2, into their atoms N and O. Within the combustion chamber, some of these nitrogen and oxygen atoms recombine with each other to form nitric oxide (NO).

So what comes out of a car's tail pipe is not a very pretty combination: a mixture of partially burned, partially oxidized hydrocarbons and nitric oxide. The amount of NO coming out of the exhaust is not critically dependent on how efficient the combustion process is within the combustion chamber. Whether it's a $10,000 or a $30,000 car makes no difference in how much nitric oxide is coming out its rear end. What does matter is the temperature within the very hot combustion chamber. The higher the temperature, the more NO. The catalytic converters required on exhaust systems in the 1970s oxidized some of the partially oxidized hydrocarbons and CO to make more CO_2 and water vapor. But they did nothing to reduce the amount of NO coming out in the exhaust.

Remember those partially oxidized reactive hydrocarbon radicals? The ones we called RO_2? Well, under favorable conditions, those radicals will react with the NO in the air, which will make another partially oxidized hydrocarbon radical and nitrogen dioxide (NO_2). This reaction looks like this:

$$RO_2 + NO \rightarrow RO + NO_2$$

where RO is another reactive hydrocarbon radical. This process then touches off another set of chemical reactions which becomes very difficult to describe and which is still not completely understood. What is understood is that the carbon atoms that comprise the RO and RO_2 radicals are dedicated little things, finding any pathway, overcoming any hardship to eventually achieve their just due: carbon dioxide. Because every carbon atom in fossil fuel eventually becomes a CO_2 molecule, it is relatively easy to compute how much CO_2 comes from fossil fuel usage. A more difficult task is to estimate how much carbon monoxide (CO) or any other intermediate product of combustion comes from burning hydrocarbons. Such an estimate must include the type of fuel burned and a knowledge of how efficiently the combustion process takes place.

Meanwhile, the nitrogen dioxide molecule that was formed in the above process by the reaction between the NO coming out of the tail pipe and the hydrocarbon fragment plays an important role in the generation of ozone. Nitrogen dioxide will break apart by absorbing visible light in the blue part of the electromagnetic spectrum. That's why the smog in Los Angeles appears brown, which is what happens when the white light that NO_2 absorbs is reemitted into the atmosphere with all the colors except blue. Thus:

$$NO_2 + \text{photon (blue light)} \rightarrow NO + O$$

This oxygen atom readily finds another oxygen molecule and a neutral molecule to make ozone:

$$O + O_2 + M \rightarrow O_3 + M$$

THE OZONE PARADOX

We mentioned in the last chapter that a fleet of supersonic transports (SSTs) injecting oxides of nitrogen into the atmosphere would destroy ozone. Now it looks as if NO coming out of cars will make ozone. Isn't this contradictory? Yes, and it is this contradiction that managed to confuse the general public for many years. The explanation amounts to an ozone paradox. The key to the confusion is the last reaction, $O + O_2 + M \rightarrow O_3 + M$. Once the oxygen atom is formed, it will regenerate ozone if, according to the above reaction, it can latch on to another oxygen molecule at the same instant that another neutral molecule (shown by M in the above equation) bumps into the O_3 molecule that has just been formed. Without the simultaneous presence of this third party, the O and the O_2 would separate again.

Thus, to have the reaction proceed efficiently, there must be lots of O_2's and M's around. Otherwise, the oxygen atom will react with something else in the atmosphere. In the rarefied air of the stratosphere, the density of the air is typically only 1 percent what it is at the surface, and the likelihood of the recombination of an oxygen atom with an oxygen molecule is relatively small. The chances of this reaction taking place in the atmosphere is, in a first approximation, proportional to the product of the amount of O_2 and M present. In the lower atmosphere, where the density of the air is approximately 100 times greater than in the stratosphere, the chances of an oxygen atom's recombining with an oxygen molecule is 10,000 (100 times 100) times more likely than in the stratosphere.

The crucial question, then, is where exactly is the threshold where NO starts destroying ozone rather than generating ozone? The answer appears to be somewhere in the stratosphere. Conventional airplanes flying in the upper troposphere end up making ozone, or to phrase it differently, they are making smog at 35,000 feet. Some theoretical calculations even show that SSTs flying in the lower stratosphere actually may make, rather than destroy, ozone. The higher in the stratosphere NO is directly injected, the more likely it is to destroy ozone.

RETURN TO EARTH

So the ingredients necessary to produce ozone in the lower atmosphere are hydrocarbons (a good source being gasoline), nitric oxide (a good source being the internal combustion engine), and sunlight. It's easy to see why the Los Angeles area, where the automobile is so ubiquitous and the sunshine so plentiful, is prone to ozone formation. In Figure 10, the average concentrations of nitric oxide (NO), nitrogen dioxide (NO_2), and ozone (O_3) are shown for a particular day in a downtown urban area. These hourly averaged measurements show how nitric oxide peaks in the morning, coincident with heavy morning-rush-hour traffic. Nitric oxide is called a *primary pollutant* because it is emitted directly

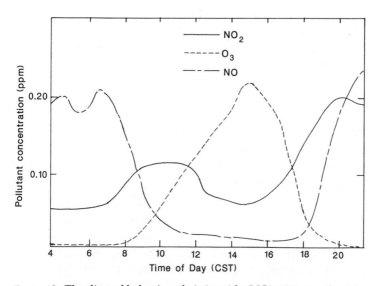

Figure 10. The diurnal behavior of nitric oxide (NO), nitrogen dioxide (NO$_2$), and ozone (O$_3$). Nitric oxide is a primary pollutant and reaches its highest morning concentration in conjunction with rush hour traffic. Atmospheric reactions convert the nitric oxide to nitrogen dioxide, which reaches its highest daytime concentration later in the morning. As photochemical reactions continue, ozone is produced, which typically reaches its highest concentration in the afternoon. This particular data set represents an average of four surface-monitoring stations in St. Louis.

into the atmosphere. Carbon monoxide, hydrocarbons, sulfur dioxide, and soot are likewise primary pollutants.

The nitrogen dioxide peaks later in the morning and ozone peaks later in the afternoon. These two trace gases are called *secondary pollutants* because they are not emitted directly but are produced in the atmosphere as a result of photochemistry acting on primary pollutants. The peaks shown are consistent with the chemistry summarized earlier. Reaction times for the formation of ozone in the urban atmosphere are on the order of several hours. The emissions from the evening rush hour do not result in ozone formation because there is no sunlight.

In Los Angeles, the situation is compounded by the fact that air is often trapped in the lowest few hundred meters by the presence of a thermal inversion layer, as mentioned earlier. This meteorological condition allows the pollutants to remain trapped and produces high concentrations of ozone. As we noted, extremely high concentrations of ozone greater than 0.5 parts per million were first observed in the LA area in the early 1950s as southern California's population boomed, and the use of the automobile also boomed.

SMALL SCALE VS. LARGE SCALE

By the early 1970s, the stage was set for an interesting sparring match in the field of atmospheric chemistry. More than two decades after scientists had begun researching why the air around Los Angeles was so bad, the scientific community that had been concentrating on the air pollution problem had come to realize that photochemical reactions were the key to understanding ozone generation. But such reactions were important only in such peculiar places as Los Angeles. Or were they?

Through the 1950s and 1960s, conventional wisdom said that ozone formation at such a relatively localized sites as the LA basin could not have a significant impact on the global ozone budget throughout the entire troposphere. After all, the high ozone concentrations found in southern California extended to an altitude of only a few tenths of a mile, and the expanse of the superhighways covered only a few hundred square miles. Both of these domains seemed almost infinitesimally small when compared to a layer in the stratosphere of high ozone concentrations that was more than 12 miles deep and encompassed the entire planet, an area of hundreds of millions of square miles.

Members of the atmospheric chemistry community are still debating this theory of the origin of tropospheric ozone. But observational evidence is beginning to mount that most of the ozone in the background troposphere, once believed to be natural in origin, may be the result of human activity, at least in the northern hemisphere. We already know that the air in our larger urban centers is bad. We know that when the Environmental Protection Agency issues "ozone alerts" we're to stay inside. Maybe we figure that's the price for living in the big city, though many urbanites would find that price unacceptable and unnecessary. But the ozone that is the focus of this book is the ozone that exists in concentrations that are now commonly found not in the urban centers but in the rural regions of the entire northern hemisphere. The summer of 1988 provided some hints as to what might be the norm of the future. As we mentioned, residents of Maine were cautioned for the first time to limit their exercise outdoors because of high ozone concentrations. In the Appalachian Mountains of southwest Virginia, monitoring stations measured ozone concentrations above the National Ambient Air Quality Standard (NAAQS) for the first time ever. Despite the sporadic nature of these occurrences, the fact that they had never before occurred gives strong evidence that the pollution must have originated far from where it was measured.

Although the ozone concentrations found in such rural areas most often are below levels at which pollution alerts would be put into effect, the levels present are considerably higher than the concentrations recorded only a few decades ago. Between 1965 and 1985 carbon dioxide concentrations increased by 6 percent, and background ozone concentrations went up more than twice as fast. The data show that, unlike CO_2, however, ozone has increased only in the northern hemisphere, which gives us some clue that the story behind this increase in ozone in the lower atmosphere is considerably more involved than the CO_2 story.

Nonetheless, the story of the increase in tropospheric ozone may be one of the most intriguing in the scenario of global change. In addition, global smog is very likely among the most significant stories because it affects the human population and life on this planet more immediately and urgently than either the increase of carbon dioxide or the depletion of the ozone layer in the stratosphere.

NATURAL GAS, UNNATURAL CONCENTRATIONS

Interestingly, the ozone story began where most of the stories about other trace gases in the atmosphere began—in a published paper. This particular paper appeared in *Science* magazine in 1971. It was written by Hiram "Chip" Levy II (see Figure 11), who at the time had recently received a Ph.D in chemistry from Harvard and was a postdoctoral fellow at the Smithsonian Astrophysical Observatory in Cambridge, Massachusetts. In the classical sense, Levy did some "back-of-the-envelope" calculations that formed the backbone of tropospheric photochemistry as we know it today. Using a very crude representation of the atmosphere (i.e., temperature, air density, amount of water vapor, ozone, and the amount of ultraviolet radiation that should be reaching the lower atmosphere), Levy's calculations suggested that our old friend the hydroxyl radical (OH) exist in the background troposphere in sufficient quantities to initiate important photochemical reactions that had previously been believed to exist only in the "polluted" atmosphere of the large urban areas.

Levy's calculations focused on methane (CH_4), the second most abundant trace gas in the atmosphere (after carbon dioxide) and the only other trace gas that is ubiquitously found in the background atmosphere at concentrations of more than 1 part per million (ppm). The difference between methane, which exists in concentrations of about 1.5 ppm (circa 1970), and CO_2, which exists at more than 350 ppm, is that methane is removed primarily by photochemical processes in the atmosphere whereas CO_2 is photochemically nonreactive in the troposphere.

Figure 11. Hiram "Chip" Levy II (1940–), an atmospheric scientist at National Oceanic and Atmospheric Administration's Geophysical Fluid Dynamics Laboratory in Princeton, New Jersey. Levy published the first paper (in 1971) suggesting that photochemical reactions in the unpolluted lower atmosphere are an important component of atmospheric chemistry. (Photograph courtesy of Lois Levy.)

Levy's premise, then, was that there should be enough of those hydroxyl radicals in the atmosphere to create significant concentrations of formaldehyde (CH_2O) in the background atmosphere. The reason for the formaldehyde is a process Levy called "methane oxidation," the formaldehyde being the by-product of the oxidation of methane. Once the formaldehyde is formed, it is also influenced by photochemical reactions and ends up producing carbon monoxide (CO) and hydrogen (H_2) as end products. Since methane is a natural component of the atmosphere, Levy's calculations suggested that carbon monoxide, an end product of methane oxidation, is likewise a natural component of the unpolluted atmosphere.

Thus, the early 1970s saw a series of theoretical calculations, based on the presumption of methane oxidation, appear in the scientific literature. An extension of the Levy hypothesis was published in the *Journal of Geophysical Research* in 1973 by a group of Harvard researchers (Steve Wofsy, Jack McConnell, and Michael McElroy), who concluded that the source of carbon monoxide resulting from the oxidation of methane in the natural atmosphere was approximately 10 times more than the source of CO that was produced from fossil fuel combustion. This theory of the origin of carbon monoxide perpetuated itself through the 1970s because of the lack of measurements of the gas which could have been used to validate or refute the hypothesis. If the Harvard hypothesis were true, then equal amounts of carbon monoxide should be found in both the northern and southern hemispheres since methane concentrations were believed to be approximately the same in both hemispheres.

Even though the Harvard researchers were aware of Seiler's measurements and another set of ship-based measurements showing the hemispheres' differences in CO concentrations, they still professed that methane oxidation was the dominant source of atmospheric carbon monoxide. Furthermore, the National Academy of Sciences published a comprehensive report in 1977 which reiterated the belief that human activity was an insignificant source of atmospheric CO.

Yet in the same report the academy stated that industrialized activity released approximately 359 million tons of carbon monoxide into the atmosphere every year. This may seem like a lot, but it was only a drop compared to the theoretical calculations at the time, which estimated the source of CO from methane oxidation at 2,500 million tons per year. The fact that the National Academy of Sciences put this budget for CO in its document, and yet completely discarded the importance of Seiler's findings, had important repercussions at the time. If the premise that methane oxidation completely overwhelmed all other sources of CO in the atmosphere was true, then there was very little the civilized world could do to control CO buildup in the atmosphere, since humanity's relative contribution was so small. Such a budget did little, if anything, to put pressure on such forces as the automobile industry or the power industry to restrain CO emissions for global purposes.

While the concept of methane oxidation was much in vogue in the early 1970s, two other independent theoretical studies appeared in scientific literature showing that methane oxidation plays a significant role in the formation of ozone in the unpolluted troposphere. In 1973, Bill Chameides (see Figure 12) published an article

Figure 12. William L. Chameides (1949–), an atmospheric scientist at the Georgia Institute of Technology. In a series of research articles based on his doctoral research at Yale University in the early 1970s, Chameides proposed a photochemical theory for the existence of tropospheric ozone in the background atmosphere. These controversial papers were the center of much scientific debate through the remainder of the 1970s. (Photograph courtesy of the Georgia Institute of Technology.)

in the *Journal of Geophysical Research* that summarized a series of calculations suggesting that ozone formation from methane oxidation is an important, perhaps even the dominant, source of ozone in the troposphere. These calculations were originally part of Chameides's doctoral dissertation from Yale University.

At about the same time, Paul Crutzen (see Figure 13) performed a similar set of calculations that appeared in *Tellus* in 1974, and that also showed that methane oxidation as a source of ozone might be significant in the unpolluted troposphere. Both Chameides and Crutzen calculated approximately the same numbers, and their articles (in 1976 Chameides published a second article based on his dissertation) touched off quite a debate in scientific literature.

The controversy revolves around two opposing theories of the origin of tropospheric ozone: "mixing" and "photochemistry." Chameides's premise was contrary to the popular notion of the time, a notion that had been accepted since the 1940s, which stated that chemistry played an important role in ozone formation, but only in the stratosphere. Any ozone found in the troposphere (except for the high concentrations observed in the large cities) must have had its origin in the stratosphere since this reservoir of ozone was so large and so ubiquitous. How the ozone was "mixed" down from the stratosphere into the troposphere was not completely understood, but it was known that there are several physical mechanisms that can transport stratospheric air into the lower atmosphere.

In fact, the most comprehensive set of ozonesonde measurements ever recorded were obtained in the 1960s, but not because of an interest in ozone. What

Figure 13. Paul J. Crutzen (1933–), a Dutch meteorologist. Crutzen's doctoral research at the University of Stockholm and his postdoctoral research at Oxford University in the late 1960s led to concern about the depletion of the ozone layer by supersonic aircraft flying in the stratosphere. In 1974, he also published a scientific paper pointing out the importance of photochemistry in the background atmosphere and its implications for the global tropospheric ozone budget. He served as director of the Air Chemistry and Aeronomy Division at the National Center for Atmospheric Research in Boulder, Colorado, in the late 1970s until he became director of the Max Planck Institute for Chemistry in Mainz, West Germany, in 1980. (Photograph courtesy of the National Center for Atmospheric Research of the National Science Foundation.)

scientists were interested in was the mixing between the stratosphere and troposphere, and they used ozone as a marker to distinguish the two layers of the atmosphere. Their interest was fueled by historical events of the time: the Chinese and French, in the late 1950s and early 1960s, were conducting nuclear bomb tests in the atmosphere. Much of the radioactive debris from these explosions settled in the lower stratosphere.

Although much of the radioactive material eventually settled lower and lower in the stratosphere, sporadic events could quickly transport large amounts of stratospheric air into the troposphere. Through a series of aircraft flights that measured both radioactivity and ozone, Edwin Danielsen produced a meticulous set of elegant analyses for several case studies in the early 1960s, in which ozone was transported downward through a process called *tropopause folding*. Danielsen devoted his scientific career to understanding the processes that controls transport between the stratosphere and the troposphere. This topic formed the core of his dissertation from the University of Washington in the late 1950s until his retirement from NASA's Ames Research Center in California in the late 1980s. At the time of the Chameides papers, Danielsen was a group leader at the National Center for Atmospheric Research in Boulder, Colorado.

Danielsen discovered the presence of such "folding" events during the development of large-scale storms, those that produce the nasty weather associated with late-winter and early-spring blizzards in the upper Midwest, or even tornadoes farther to the south. Danielsen's analyses showed that the jet stream, the powerful "river" of high-velocity air that moves our surface weather systems across the continent, can descend from the boundary between the troposphere and the stratosphere (typically about 40,000 feet at midlatitudes) to much lower altitudes, during which time air containing high levels of ozone and radioactivity from the lower stratosphere is transported rapidly into the middle and lower troposphere. By knowing how much ozone is typically transported out of the stratosphere by each storm, an estimate of the amount of ozone being transported over the course of a year could be calculated by multiplying that amount by the number of storms that occurred each year. From such a calculation, it was estimated that approximately 500 million tons of ozone come out of the stratosphere each year.

Despite Danielsen's detailed analyses of some specific examples of stratosphere–troposphere exchange, some of his colleagues didn't totally agree that the formation of storms (cyclones) is the primary means by which ozone came into the troposphere from higher altitudes. One such critic was Elmar Reiter, a professor of atmospheric science at Colorado State University in Fort Collins. Reiter maintained that the process of air exchange between the two regions is a more gradual process and that the relatively few sporadic events described by Danielsen comprise only one part of the story. Despite the differences in the mechanism, however, Reiter also concluded that the amount of ozone coming down from the stratosphere is on the order of 500 million tons per year.

In a series of completely independent studies done earlier (in the 1950s and 1960s), a group of German scientists had continued their studies on ozone in the troposphere and were interested in finding out exactly how much ozone is destroyed at the earth's surface over the course of a year. Their observations indicated there are always relatively low values of ozone right next to the earth's surface. As measurements were made vertically away from the surface (usually from towers rising a few tens of meters high), they observed that ozone increased in concentration. The difference between the concentrations above the ground and higher up in altitude is known as the *gradient*. The gradient, they discovered, is considerably stronger over land than over either water or ice. In other words, the difference in ozone concentrations at ground level and tens of feet above ground level is generally greater over land than over water or ice. Such a vertical distribution near the earth's surface was proof that the ground is an important sink term in figuring out how much ozone there is in the entire troposphere.

In addition, the gradient seemed to vary depending upon the type of surface (e.g., desert, grassland, or crops), the time of day, and the season of the year. Thus, if the gradients of ozone near these various surfaces could be characterized for time of day and season, estimates could be made of how much ozone is deposited on the earth's surface. By taking a global inventory of the number and types of surfaces present · ⁻⁻ the world, an estimate could be made of how much ozone is lost from the atmosphere globally because of this deposition process.

Scientists took such an inventory, and the figures summed up over the entire planet came out to be about 500 million tons each year. It seemed clear that the amount of ozone coming out of the stratosphere is perfectly balanced by the amount of ozone deposited over the earth's surface. Therefore, the conventional school of thought allowed for no other source of tropospheric ozone to be of any importance in the global tropospheric ozone cycle. That's why, when Chameides presented his photochemical theory of tropospheric ozone, there was an outcry from the traditional school (the "mixers"), who had been studying the problem since the early 1950s. As far as they were concerned, Chameides's ideas bordered on scientific anarchy.

The mixers argued that the observations simply did not support the notion that photochemistry could play any significant role in the global budget of tropospheric ozone, considering the vastness of the stratospheric reservoir. Their main argument against photochemistry was that there is considerably less ozone in the tropics than at mid and high latitudes. If photochemistry were important, then it should be most important at low latitudes, where there is the most intense solar radiation.

A puzzling paradox. If photochemistry were an important component of the tropospheric ozone budget, why was it not equally important everywhere in the troposphere? One key to the answer lay in the recognition of Seiler's earlier global measurements of carbon monoxide, which showed that there are two to three times more CO in the northern hemisphere than in the southern hemisphere. The second key lay in an improvement of laboratory techniques which enabled reactions involving radicals in the atmosphere (such as OH and RO_2) to be measured with considerably more accuracy. In particular, new laboratory data obtained around 1974 showed that the speed at which OH reacts with CO is more than twice as fast as had been previously measured. These two factors had important implications for our understanding of both the global CO budget and, consequently, the global tropospheric ozone budget.

This new information completely invalidated the work performed by the group of Harvard scientists in the early 1970s. Using these new inputs, Crutzen and his colleagues showed that the chances of a CO molecule's reacting with OH is three times more likely than a methane molecule reacting with OH. Thus, since the Wofsy–McConnell–McElroy Harvard calculations showed that methane oxidation resulted in the formation of 2,500 million tons of CO, the calculations based on these new measurements forced a contribution of at least 5,000 million tons of CO to come from other sources, since OH reacting with CO would result in the destruction of approximately 7,500 million tons each year.

Even Seiler, who strongly believed that manufactured CO in the northern hemisphere dominates all other sources of CO, could come up with a contribution of only 640 million tons per year from fossil fuel usage. Even though this estimate was nearly twice the estimate of the National Academy of Sciences, it was far short of the 5,000 million tons that would have been needed if the amount of OH in the atmosphere theorized by the Harvard group had been correct.

Crutzen and his co-workers resolved this dilemma by stating that the concentrations of OH that had been calculated in all of the earlier studies must have been too high by a factor of 3 to 5. OH had never been measured in the atmosphere at the time of these calculations because it is present in such very low concentrations (of less than one part per trillion) owing to its highly reactive nature. Even with these much lower theoretical OH numbers, the CO budget could not be balanced by the sources, the main ones being fossil fuel combustion and a revised contribution from methane oxidation.

Crutzen's work appeared in *Geophysical Research Letters* in 1977. One of the primary conclusions in that paper was that there must be other very substantial sources of CO. Crutzen hypothesized that one large source of CO previously overlooked is widespread bio-mass-burning from land clearing and agricultural practices, primarily in the tropics. As director of the Atmospheric Chemistry Division at the National Center for Atmospheric Research in Boulder, Colorado, Crutzen deployed an entourage of chemical instruments aboard a research airplane to take measurements of the trace gases emitted from biomass burning in Brazil in 1979 and 1980. As he observed the enormous clouds of smoke from the Brazilian regions, Crutzen set out to learn as much as he could about the combustion of various materials. The process of how things burn dominated much of his research during these years, and one of his hypotheses soon made him somewhat of a celebrity. His hypothesis was this: An enormous quantity of smoke would be generated by the burning of cities and forests after a nuclear war. Crutzen theorized that this huge quantity of smoke would eventually impact the entire globe, resulting in a change in both the composition of the atmosphere and the temperature of the world. Thus, he introduced a new component into the nightmare of a nuclear war. A subsequent paper by Rich Turco, a stratospheric modeler who worked for a consulting firm in California, and his colleagues took Crutzen's hypothesis one step further and examined the climatic implications of all this smoke produced by a nuclear holocaust. Their calculations showed that it was possible that all of this smoke and soot released after the explosions would result in a drastic cooling of the planet. In the early and mid-1980s, this scenario became known as *nuclear winter*, for which Crutzen was recognized by *Discover* magazine as its Scientist of the Year for 1984.

But Crutzen's main interest was in other aspects of atmospheric chemistry. By the late 1970s it could be shown that CO reacts with OH in the atmosphere much faster than had previously been believed. Furthermore, it is likely that a by-product of this atmospheric photochemical process is the production of ozone in the troposphere. The first reaction proceeds:

$$CO + OH \rightarrow CO_2 + H$$

The H atom then very quickly (within a small fraction of a second) latches onto an oxygen molecule to form a reactive peroxy radical, HO_2. This HO_2 radical is analogous to the RO_2 radicals produced in the urban smog scenario discussed earlier. The key to whether or not ozone is produced in the remote atmosphere centers on what happens to the HO_2 radical. If there is a molecule of nitric oxide (NO) around, then it is likely that the following reaction takes place in the atmosphere:

$$HO_2 + NO \rightarrow NO_2 + OH$$

The nitrogen dioxide will be photolyzed by visible light and ozone will again be made, just as it is in the polluted urban environment:

$$NO_2 + \text{(visible) photon} \rightarrow NO + O$$

$$O + O_2 + M \rightarrow O_3 + M$$

The above sequence of reactions that results in ozone formation is a catalytic cycle with respect to the hydroxyl radical and nitric oxide. Both OH and NO are returned to the atmosphere so that more carbon monoxide can be oxidized to carbon *di*oxide by the hydroxyl radical, and likewise so that nitric oxide can be converted to nitrogen dioxide to make more ozone.

This reaction sequence is important because a significant fraction of all the carbon monoxide in the atmosphere that reacts with the hydroxyl radical will result in the production of ozone. Thus, since the newly revised calculations of Crutzen and his colleagues showed that 1,000–2,000 million tons of CO are removed by the OH radical each year, a sizable fraction of this reservoir of carbon monoxide now serves as a source of ozone in the atmosphere. Such a source is considerably larger than what had been previously estimated as coming from transport out of the stratosphere or destruction at the ground. Both of these estimates were on the order of 500 million tons annually. Crutzen's scenario also explains why more ozone is formed in the northern hemisphere, since most of the sources of CO, and the higher concentrations of the gas observed by Seiler, had been measured in the northern hemisphere. Also, since Seiler's CO measurements had been obtained over the ocean, far removed from local pollution sources, this set of calculations appeared to be valid for the entire troposphere, not just those regions close to urban centers, as originally believed.

PROVING (DISPROVING) A THEORY

The theory that the tropospheric ozone budget is closely linked to the carbon monoxide budget was published in *Nature* in 1978 by one of the authors of this book (Jack Fishman) and Crutzen. Fishman had studied under Crutzen's direction and had recently received his Ph.D. In 1978 he worked with Crutzen as a research associate at the National Center for Atmospheric Research in Boulder. The crux of the analysis in the *Nature* article centered on the difference that should exist between the two hemispheres. On the one hand, the *Nature* paper showed that there is 30 to 40 percent more ozone in the northern hemisphere than in the southern hemisphere. But Fishman and Crutzen also pointed out that the northern hemisphere contains three times more land mass. This latter point was crucial because it allowed for a new estimate of how much tropospheric ozone is deposited at the earth's surface, which the Germans had been studying since the 1950s.

Because ozone is destroyed so much more efficiently over land than over water or snow surfaces, the *Nature* study examined how this factor of the global budget should be different in the two hemispheres. The bottom line was this: Three to four times more ozone should be destroyed in the northern hemisphere than in the southern hemisphere. Thus, the source of ozone in the northern hemisphere has to be comparably disproportionate to the source of ozone in the southern hemisphere to account for the observed distribution.

If, as the mixers believed, all of the ozone in the troposphere comes originally from the stratosphere, then three to four times more ozone must be coming down in the northern hemisphere than is coming down in the south. Up to that point, no one had studied the stratosphere–troposphere exchange with respect to how it might be different in the two hemispheres.

Although Ed Danielsen had described in detail how, during the development of a storm, air can be brought down from the stratosphere, his observations required the proper placement of an airplane flying around and taking measurements in the vicinity of the occurring process. Those measurements were then analyzed in conjunction with other, more conventional measurements of wind, temperature, humidity, and pressure, to interpret how the aircraft data fit into the larger scale meteorological picture. The scientific value of Danielsen's work was unique because of the detail he put into his analyses to show the "whole picture" of the stratosphere–troposphere exchange. The price the scientific community had to pay for these analyses was time: they took years to complete. Thus, although his studies clearly showed that ozone comes down from the stratosphere, it was humanly possible for Danielsen to do no more than a handful of case studies through the 1960s and early 1970s. For logistical reasons all the studies had been done in the United States.

So how do we know whether or not three to four times more air is coming down from the northern stratosphere rather than from the southern stratosphere? The *Nature* study examined the frequency of storms in each hemisphere, although this analysis could serve as only a crude relative surrogate for how air comes from the stratosphere to the troposphere. Because there are fewer people living in the southern hemisphere than in the north, there are considerably fewer meteorological observations there.

A detailed study of the meteorology of the southern hemisphere was written by Harry van Loon, a South African meteorologist who had been working at the National Center for Atmospheric Research (NCAR) since the 1960s. Van Loon's analyses suggested that the number and intensity of the storms that were responsible for bringing stratospheric air downward should not be appreciably different in the two hemispheres. He certainly could not reconcile a factor difference of 3 to 4.

Thus, the *Nature* paper offered the alternative explanation, that the observed difference in the carbon monoxide concentrations, and its subsequent oxidation by the hydroxyl radical, is responsible for the hemispheric difference. But such an explanation was still hypothetical, and there were many atmospheric scientists who still didn't believe it. Furthermore, the question of the relative amounts of air coming down into each hemisphere had never even been asked before, let alone correctly answered.

One way to try to answer the question was through the use of a general circulation model (GCM). These models try to explain how air moves around the world by solving numerical equations describing the underlying physics that lead to the formation of weather. These numerical models solve these equations at fixed points (usually equally spaced) around the world and for several vertical levels in the atmosphere. Since the size and speed of computers have increased dramatically since the 1950s, the models can use more points in both the horizontal and the vertical domains. Because of the complexity of such models and the enormous amount of computer time needed to run them, there are no more than a handful of GCMs in the world.

Following the publication of the 1978 *Nature* article, scientists conducted two independent analyses of GCM computations to determine what the models would say about the differences between the two hemispheres. Louis Gidel and Mel Shapiro analyzed the results from the GCM at NCAR; a group from the Geophysical Fluid Dynamics Laboratory at Princeton University, composed of Jerry Mahlman, Chip Levy (who had been at the Smithsonian Observatory in 1975), and Bud Moxim looked at this parameter from their GCM. Both groups concluded that there is more transport of stratospheric air into the troposphere in the northern hemisphere, but that the difference is on the order of 50 to 80 percent, not the factor of 3 to 4 necessary to account for the calculated differences in the destruction rates at the earth's surface in the two hemispheres.

More significant, however, was the fact that the actual amount of stratospheric air coming down into the troposphere is less than the 500 million tons that had been estimated by the original mixing-dominated theories.

Despite the absence of proof against the carbon-monoxide–ozone relationship, firm evidence in support of the hypothesis was still lacking. In June 1978 a small session on atmospheric chemistry was held in Philadelphia at the annual meeting of the American Institute of Chemical Engineers. One of the scientists in attendance was Wolfgang Seiler from the Max Planck Institute for Chemistry in Mainz, West Germany. Seiler had just made his mark for his newly fashioned carbon-

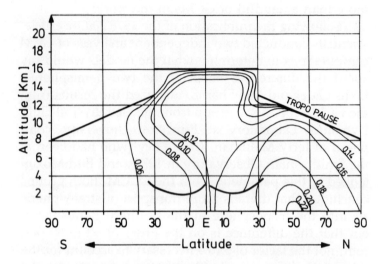

Figure 14. The global distribution of carbon monoxide obtained by Wolfgang Seiler. The heavy solid lines indicate the approximate positions of the tropopause and the height of the trade wind inversion in the tropics. The units of the isopleths (contour lines) are parts per million. Note the much higher concentrations in the northern hemisphere than in the southern hemisphere. (From W. Seiler, "The Cycle of Atmospheric CO," *Tellus*, 26, 1974, pp. 116–135. Reprinted with permission.)

monoxide-measuring instrument, the one that measures low levels of CO almost instantaneously. Since the late 1960s, Seiler had been using the instrument aboard airplanes and ships and at remote locations over long periods of time, to map the global distribution of CO and to determine its seasonal variability in the unpolluted atmosphere. In his 1974 *Tellus* paper, he had summarized his first sets of measurements, obtained through 1971 (see Figure 14).

But it was now 1978, four years after Seiler had taken his six-week journey around the world to measure carbon monoxide in the stratosphere. Between 1974 and 1978, Seiler had been most preoccupied with obtaining more CO measurements. He had not yet let the scientific community know about the details locked up in his 1974 flights. He had never really considered the importance of the strange behavior of the needles during his plane trip four years previously. But at the meeting in Philadelphia, Seiler heard a talk by Fishman, based on the CO–ozone relationship. Fishman concluded his talk by reiterating that his study was still theoretical, and that no measurements existed in the remote troposphere verifying the theory. He was also not aware of Seiler's 1974 airplane measurements. But all that would soon change after the two men finally met face to face.

CHAPTER FOUR

Traces

In April 1979 Jack Fishman arrived in Germany to begin his five-month collaboration with Wolfgang Seiler, the German scientist he had befriended during the latter's stay in America. The two scientists shared an interest in atmospheric chemistry—specifically in answering the questions of quantity (How much is there?), origin (Where does it come from?), and destination (Where does it go?) of the various gases found in our atmosphere. Though both men had devoted their professional lives to pursuing the answers to those questions, their careers had taken two distinct directions. One scientist had written his dissertation on carbon monoxide, the other on tropospheric ozone.

Sitting atop a file cabinet in Seiler's office was a small box covered with a thick layer of dust. The German host brought down the box to show to the American. As he opened the box a thick cloud of dust seemed to mock the room containing equipment used to measure minute amounts of trace gases in pristine air. In-

Figure 15. Christian Friedrich Schönbein (1799–1868), professor of chemistry at the University of Basel, Switzerland. Schönbein is credited with the discovery of ozone in 1839. He devoted much of his career to the measurement of ozone at numerous locations throughout Europe to prove that it was a natural component of the atmosphere. (Photograph from *Pioneers of Ozone Research: A Historical Survey*, Max Planck Institute for Aeronomy, Katlenburg-Lindau, West Germany. Reprinted with permission.)

side the box were rolls of paper, similar to toilet paper but smaller. These were called *stripcharts*, and on them were recorded data that could provide valuable information about Fishman's iconoclastic theory as to the origins of tropospheric ozone. In this dusty box atop a file cabinet halfway around the world was the key to proving his hypothesis. Or so Fishman hoped. He had gambled on his friend's good faith. He had come all this distance and committed himself to half a year of study, to discover once and for all if his theory was right.

MEASURING OZONE

The word *ozone* comes from the Greek *ozein*, which means "to smell." Probably the name of this gas originated from early laboratory studies when ozone was first discovered because of its distinctive acrid smell. The German scientist Christian Friedrich Schönbein (see Figure 15) is credited with ozone's discovery in 1839, while he was a professor at the University of Basel in Switzerland. Rather than discovering a completely new substance, Schönberg simply stumbled upon a different form of oxygen. The common form of oxygen normally found in the atmosphere is a diatomic (two-atom) molecule, and what Schönbein discovered was a much more reactive state of oxygen, one comprised of three oxygen atoms instead of two. He called this form of oxygen *ozone*.

But what does it mean to be "reactive"? Is it like a disease? Not necessarily. A reactive molecule is similar to a hyperactive child: it doesn't like to sit still. The reason ozone is so reactive is that as a triatomic form of oxygen it yearns to re-form back to its more natural diatomic state. It will readily give up its third oxygen atom to whatever it comes in contact with; thus it is a powerful oxidizer. This oxidizing power is what makes ozone so damaging to so many things. It kills bacteria and makes nylon stockings run. The extra oxygen atom that ozone readily gives to the molecules of living tissue, nylon, rubber, or anything often transforms these molecules into new molecules that no longer serve the function they were originally intended for. As a result, rubber loses its elasticity, nylon loses its strength, and living tissues eventually die.

One of the goals of Schönbein's research was to show that ozone is a permanent and natural component of the atmosphere. He devised a method to measure ozone in the atmosphere that was capable of measuring very low levels simply and easily. The method used soon became known as *Schönbein paper* and involved the simple process of saturating a strip of paper with potassium iodide and then allowing it to dry. In the presence of ozone, the potassium iodide oxidizes and is converted to potassium iodate. In the process of this conversion the paper changes color to various hues of blue. The more ozone present in the air, the bluer the paper. Schönbein calibrated the amount of color change into a measurement standard called *Schönbein units*, which allowed Schönbein, or any other scientist, to put a new piece of Schönbein paper out each day and measure the relative amount of ozone in the atmosphere.

Although the methods of measurement have been modified over the years, scientists continued to use potassium iodide to measure ozone for more than a century. One modification involved pumping ambient air through a potassium iodide solution and measuring the amount of iodide that was converted to iodate. When iodide is converted to iodate (an oxidized form of iodine in solution), a minute electrical current is created in potassium iodide solution, similar to what occurs in a battery. So one way to measure ozone was to measure the electric current in the solution. Such technology was readily available, and soon scientists were measuring ozone with even more speed and accuracy. The reaction within the iodide solution occurred in a matter of seconds and the amount of electric current was easily quantifiable. This method, known as the *wet method*, was the predominant way scientists measured ozone until the 1960s, when other methods using newer technology made the wet method obsolete. One problem with the wet method was that other chemicals in the atmosphere will also oxidize iodide to iodate. At the same time, some trace gases could possibly interfere with this chemical reaction. The most common of these interfering trace gases is sulfur dioxide (SO_2), a pollutant that is a by-product of coal combustion. If sulfur dioxide is also present with the potassium iodide in the solution, ozone will likewise react with the SO_2, but an electric current is not generated. Therefore, the presence of SO_2 could cause the scientific observer to *underestimate* the amount of ozone in the ambient air. Hence, scientists began looking around for other methods of ozone measurement that would eliminate the sulfur dioxide interference.

Gas Chromatography

Because of its eagerness to give back its loosely attached third oxygen atom, ozone reacts with so many things so quickly that it is relatively easy to quantify by measuring a parameter that is associated with a particular chemical reaction such as electric current, heat, or light. It's just a matter of finding a reaction that is controllable and therefore quantifiable. One impractical technique might involve seeing how many pairs of nylon stockings are ruined on a particular day because of the presence of ozone: the more ozone, the more holes in the nylons. Though such a technique would be difficult to calibrate, given enough time and stockings we could probably measure ozone concentrations with relative accuracy (relative within the time frame of the measurements). But such a technique would not be very practical. It would be more practical to have ozone react with a piece of colored paper and to measure the amount of color change, or to bubble it through a solution and to measure the amount of electricity generated.

Even more challenging than measuring ozone is measuring other, less reactive trace gases in the atmosphere, those that do not react quickly with materials they come in contact with. So even though carbon monoxide is more abundant than ozone in the lower atmosphere, techniques for measuring it require considerably more sophisticated equipment, especially because of the low concentrations (often less than 1/10 of a part per million) commonly found in the background atmosphere. In urban areas it is another story, for there concentrations often exceed several parts per million, and measuring such large concentrations is relatively easy. It was the difficulty of measuring very low concentrations of trace gases that led to the development of a technique called *gas chromatography*. In gas chromatography, the compounds in an air mixture migrate at differing speeds when carried along by an inert gas that has been packed or treated in a certain way. The method was first suggested in 1941 by the British chemists A. T. James and A. J. P. Martin. It took about 15 years to develop, and the first commercial gas chromatograph appeared on the market in 1955.

Here's how it works: A "slug" of air is allowed to flow very slowly through a series of filters and then through a very long, thin coiled tube. These filters, called *columns*, remove many of the trace gases not of interest, or those which would interfere with the actual measurement of the gas being measured. The remaining components of the air will then separate themselves, usually according to their relative weights. For example, molecular hydrogen is the lightest gas in the atmosphere, with a molecular weight of 2. It will pass through the column most quickly, followed by such gases as methane (molecular weight of 16) and carbon monoxide (28). As each of these gases passes through the gas chromatograph, a special process is used to measure them. One commonly used process involves putting a special flame at the end of the line. As each of the gases is burned, the flame becomes brighter, and the brightness of the flame is then measured.

Gas chromatography is not an absolute measure and must be carefully calibrated. To calibrate it, the scientist or technician must know in advance which gases will pass through the column and how long it takes each one to make its journey before it arrives at the detector.

For example, using a carefully controlled flow rate, hydrogen may take 30 seconds, methane 60 seconds, and carbon monoxide 80 seconds. Once these gases are subjected to the detector at the end of the line, their responses must also be calibrated. For example, 1 part per million of hydrogen may burn brighter than 2 ppm of methane. Thus, before beginning the test measurements, a known mixture of the same gases to be measured later is run through the gas chromatograph and the responses of the gases in the ambient sample are compared.

The preparation of calibration standards for use in atmospheric chemistry is a tedious process. To start, the scientist must obtain a pure mixture of the trace gas he or she hopes to measure. The gases are generally available from commercial manufacturers after having been certified pure by the National Bureau of Standards. Rarely is a gas from a manufacturer certified as being "100 percent pure." Most often the gas in the "pure" cylinder is certified to be more than 99 percent pure. The scientist then takes one part of the pure gas and mixes it with 1,000 parts (for example) of a certified pure inert gas such as nitrogen. This procedure creates a calibration mixture of 1,000 ppm. The scientist then takes the mixture and carefully dilutes it again and possibly even several more times to produce a standard in the range of several parts per million to several parts per billion (ppb).

Once such a mixture is at the concentration desired, it can then be used as a calibration standard. But the scientist must test it periodically over several months to a year to make sure that the concentration remains stable within the gas cylinder. It is not uncommon for the inside of the cylinder to be contaminated with some impurity that may react in the cylinder with the trace gas of interest. If the concentration within the cylinder remains stable for several months, then it is used as the primary standard. After the preparation of such a standard, the scientist can go ahead and analyze the air sample.

What is actually analyzed in the gas chromatograph is a plot of brightness (or voltage, or any other measure that can be accurately quantified) within a given time frame. It's not uncommon for some of the peaks of brightness or high voltage that show up as an air sample permeates the column to remain unidentified because a controlled sample of the gas has never been run through the gas chromatograph previously. When an unidentified peak shows up, the scientist often becomes a sleuth to discover what unknown trace gas she or he has just measured for the first time in the ambient atmosphere. Many times such trace gases are never clearly identified, and other times the scientist discovers that the air sample has been contaminated—possibly by something as overlooked as the mosquito repellent the technician wore when opening the valve to scoop up some pristine air in the Florida Everglades. Such samples must be either thrown out or interpreted with the utmost caution.

The most common measuring technique using gas chromatography is to collect a sample of air in a container and bring it back to the laboratory for analysis. Normally, these containers are made of glass or stainless steel and have been emptied at the lab to make room for the gas to be studied. The air intake valve is then shut off. In the field the valve is opened, and the ambient air rushes into the evacuated chamber until the container is filled. The valve is shut and sealed until the sample has been analyzed back at the lab. Extreme care has to be taken to make sure the container is perfectly clean and that the material of the container will not change the composition of the air inside it. Certain materials have a tendency to absorb carbon monoxide, for example, if they remain in contact with it for too long. To run a sample through the gas chromatograph takes a few minutes, after which the plumbing (the tubes and valves through which the gas travels) has to be flushed with an inert gas such as nitrogen or argon. After that, a calibration gas has to be run through the system, followed by yet another flushing. Altogether, a typical gas chromatograph analysis takes several minutes for almost all the trace gases at background concentrations. Ozone, however, is the exception.

BUILDING A NEW CO INSTRUMENT

Back in the 1960s Wolfgang Seiler began graduate studies at Johannes Gutenberg University in Mainz, West Germany. His major professor, equivalent to what Americans would call a faculty adviser, was Christian Junge, who had been studying the composition of the atmosphere since the 1950s. In fact, Junge's pioneering work had earned him the honorary title of "the father of atmospheric chemistry," primarily because of his publishing the first comprehensive book on the subject in 1963. Junge's studies focused on the atmospheric cycles of many trace gases. He had realized that it was important to obtain as many measurements as possible in as many places around the world if the new science of atmospheric chemistry was to make any progress. So Junge was certainly interested in instrumentation that was light and portable and could be transported anywhere in the world.

Meanwhile, as part of his dissertation work, Wolfgang Seiler had succeeded in developing a new instrument to measure carbon monoxide. It was small and easily transportable. Also, the technique Seiler developed was an "absolute" technique, meaning it didn't have to be recalibrated after every measurement as did the gas chromatograph. Not only that, but Seiler's new instrument measured CO within a matter of seconds as air samples were fed through the sampling inlets continuously. Here was an instrument that could measure low concentrations of carbon monoxide anywhere in the world and that was fast and accurate and easy to operate.

Through the late 1960s and early 1970s Seiler collected an ensemble of carbon monoxide measurements from all over the world, using ships and airplanes as his primary sampling platforms. In a way he was like the ancient explorers, Magellan and Balboa, charting new territory. Only this territory was invisible and existed in the atmosphere. What Seiler was doing was mapping the world's atmosphere, charting the places according to the amounts of carbon monoxide present. His classic paper, published in the journal *Tellus* in 1974, was based on his findings prior to 1971. The fruit of his labor was the most comprehensive set of measurements, showing that much more carbon monoxide is present in the remote regions of the northern hemisphere than in the southern hemisphere.

After earning a Ph.D. Seiler stayed in Mainz at the Max Planck Institute for Chemistry, located on the campus of Johannes Gutenberg University. During his studies, Seiler had worked closely with the technical staff at the Institute, and it was their unique electronic and mechanical expertise that helped make his instrument work. In fact, under the direction of Seiler's adviser, Christian Junge, the Max Planck Institute had become the leading institution for basic research in atmospheric chemistry during the 1960s and 1970s, and Seiler's new measuring instrument was the kind of progress the institute was becoming known for. Seiler's instrument was based on the principle that carbon monoxide reacts uniquely with mercury oxide, but only at high temperatures. The mercury oxide is converted to mercury vapor, which gives off a characteristic blue light (similar to the mercury lamps now illuminating many of our large cities) when put through an electric current. Seiler's instrument measured the amount of light produced as the mercury oxide was converted to mercury. In turn, the amount of mercury produced was limited to the amount of carbon monoxide in the air.

THE CARBON-MONOXIDE–OZONE CONNECTION

As the science of atmospheric chemistry grew, gas chromatography developed also, to the point where such instruments could operate on board an airplane or a ship. Thus there was no longer a need to store air in clean containers to await analysis at the laboratory. By means of onboard instruments, vertical profiles of trace gases (i.e., measurements taken at differing altitudes) in the atmosphere could be obtained every 10 to 15 minutes, or with a vertical separation of a few thousand feet as the airplane spiraled its way up or down. By means of gas chromatography, carbon monoxide measurements could typically be obtained at as many as five levels in the atmosphere, confirming the conventional wisdom that all carbon monoxide exists in larger amounts near the surface of the earth.

During Seiler's six-week trip in 1974, he used his new mercury-oxide-based instrument to record carbon monoxide concentrations at various locations and altitudes. It was an exhausting survey, but the results were worth it. What made Seiler's instrument different was that it could obtain vertical profiles of carbon monoxide with the type of altitude resolution shown in Figure 16. Furthermore, Seiler had obtained coincidental vertical profiles of ozone in the troposphere and lower stratosphere, as shown in Figure 16. If the same CO vertical profile had been measured with only six or seven points at various altitudes, the very thin layers that contained relatively high concentrations of carbon monoxide (such as those seen at 3.1 kilometers and 3.8 kilometers in Figure 16) would not have been observed. If the altitudes at which high concentrations of carbon monoxide are the same as those at which high concentrations of ozone are present, then the two trace gases are said to be positively correlated. Such a positive relationship in this particular pair of profiles was unique at that time for global-scale atmospheric chemistry, since the average concentrations of both carbon monoxide and ozone are representative of background concentrations. It would not have been surprising to find such an in-phase relationship between the

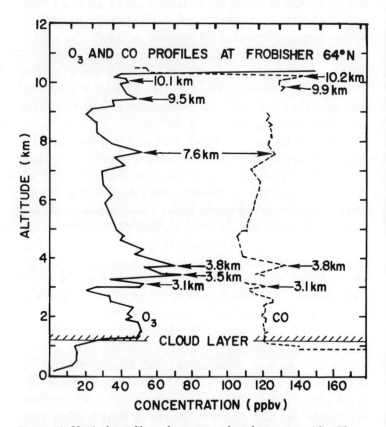

Figure 16. Vertical profiles of ozone and carbon monoxide. These measurements were obtained in July 1974 with instruments aboard an airplane during descent into Frobisher Bay, Canada, which is north of Hudson Bay, thousands of miles away from any significant urban pollution source. The coincident high values of ozone and carbon monoxide at 3.1, 3.8, and 7.6 kilometers were the first measurements to show the positive correlation between these two trace gases in the remote unpolluted atmosphere.

two trace gases if the samples had been obtained just downwind of an urban complex, since polluted air has higher concentrations of both CO and ozone. In addition, Seiler purposely tried to avoid such urban areas during his six-week expedition. These particular profiles were obtained north of Hudson Bay in northern Canada, more than a thousand miles away from any significant urban area.

Prior to the profiles shown in Figure 16, scientists interpreted the vertical distribution of ozone and carbon monoxide in the atmosphere by examining one measurement near the ground (where, in Figure 16, CO is more than 180 ppbv (parts per billion, by volume) and ozone is less than 10 ppbv), one measurement in the lower stratosphere (where CO is less than 50 ppbv and ozone is more than 120 ppbv), and only a few points in between, where the average CO concentration is ~100 ppbv and the average ozone concentration is ~40 ppbv. Thus, by looking at only a few measurements of carbon monoxide and ozone in the atmosphere, a scientist would conclude that the two trace gases were negatively correlated—the opposite of the conclusion observed when the fine structure of the measurements was compared. And Seiler's instrument was the only one capable of providing such detailed vertical structure.

Despite the fact that these measurements were taken in 1974, during the following years Seiler had neither the time nor the staff to reduce the data from those flights. Not until he heard Fishman talk about his theoretical calculations linking tropospheric ozone and carbon monoxide in June of 1978 at the Philadelphia meeting did Seiler realize he might have something useful sitting atop his file cabinet back in Germany. At the time of the meeting, Seiler was enroute to Boulder, Colorado, where he had been invited to spend six months working at the National Center for Atmospheric Research with Paul Crutzen. While in Boulder, Seiler and Fishman became good friends. Seiler told Fishman that he had a set of data which could elucidate the premise recently published by Fishman and Crutzen in *Nature*, and Seiler invited Fishman to come to Mainz and examine the set of measurements that could prove or disprove his theoretical work.

By April 1979, when Fishman arrived in Mainz to begin his five-month collaborative effort with Seiler, Seiler's data had been gathering dust for almost five years. After Seiler took down the stripcharts from atop his cabinet, he unrolled the first one and showed it to Fishman. The rolls contained what appeared to be squiggly lines, each a different color, making a pattern down the length of paper. To an untrained eye, the charts resembled the electrocardiogram that is routinely given at a doctor's office. But Fishman understood what all the lines and colors pointed to. Every few inches the lines shifted dramatically, showing that a calibration had been recorded, or that the sensitivity of the instrument had been altered. Every so often, Fishman detected a German note with an arrow pointing to one of the peaks or valleys in the squiggly record.

By today's standards, the method by which the data were stored was primitive, so it took Fishman nearly three months just to convert the traces on the stripcharts to meaningful concentrations of carbon monoxide and ozone. Seiler had bought a small computer and a digitizer several years earlier to analyze the data, but the technician he had hired to do the work had quit to make more money in private industry, a common dilemma in the research field. With the digitizer, Fishman could trace over the stripcharts with a special electronic pen and quantify every bump and wiggle along the trace.

All together, Seiler had taken nearly 40 flights during his six-week field experiment, with each flight obtaining four to six hours of measurements. During ascent and descent, Seiler kept track of altitude by having the pilot shout out the altimeter reading at exact one-minute intervals. These altitudes were recorded by him and kept in a logbook which was then used to derive the altitudes of the measurements shown in Figure 16. If Seiler wanted to duplicate the experiment today, he would simply send all of the above information directly to a computer aboard the plane. Such data are routinely recorded, and today's scientist can take a small hard disk with him and plot his results the same day. Depending on how much money, time, and talent a scientist has, the data can be displayed in real time on the airplane—sometimes even in color—and can be analyzed with the help of the same computer.

By the time Fishman returned to the United States in September, the data were in good enough shape to be described and submitted for publication in the scientific literature. Between 1980 and 1983, Fishman and Seiler published three papers describing their research, one in *Tellus* and two in the *Journal of Geophysical Research*. In the last of the three papers, they showed that there are specific regions in the atmosphere where ozone and carbon monoxide are positively correlated, as well as other regions where the two gases are anticorrelated (i.e., negatively correlated). According to their findings, the two gases are most strongly positively correlated throughout most of the troposphere in the northern midlatitudes, a finding consistent with the Fishman–Crutzen hypothesis published in the *Nature* paper. However, Fishman and Seiler's paper went on to show that most of the measurements were anticorrelated in the upper troposphere, within a mile or two of the tropopause (the boundary between the troposphere and the stratosphere), a finding consistent with the older school of thought supporting the mixing origin of tropospheric ozone. In the southern hemisphere the two scientists found both positive and negative correlations, but the positive correlations were not nearly as definitive as the ones they observed in the northern hemisphere.

LOOKING FOR TROPOSPHERIC OZONE TRENDS

Despite the relationship between ozone and carbon monoxide that Seiler and Fishman presented, other scientists expressed doubt. They argued that Seiler's measurements did not definitively prove that ozone concentrations must be increasing in the troposphere because of human activity. The most viable method of determining the ozone content of the troposphere and knowing whether it was increasing was to take measurements over a long period of time. But how good were the available ozone measurements? And how representative of the entire world or even of an entire hemisphere were measurements taken at a particular location? Such questions made it difficult to determine clearly whether or not ozone was increasing throughout the troposphere.

In the early 1980s Jennifer Logan, a research scientist at Harvard, undertook the tedious task of trying to determine whether or not ozone is increasing in the troposphere. For her sources she obtained a computer tape from the World Ozone Data Center, a clearinghouse for the coordination of all ozone data, run by Canada's Atmospheric Environment Service, located outside Toronto. As a first step, Logan had to determine just how accurate the ozone data reported to the Canadian center were. As she dug into them, she concluded that the ozonesonde data were particularly susceptible to errors. She determined that there were three distinct sources for errors in the ozonesonde measurements: (1) sometimes the pumps bringing the air into the measurement chamber did not work as efficiently in the rarefied stratosphere as they should; (2) the balloons burst at relatively low altitudes and did not obtain measurements in the middle stratosphere; and (3) sometimes the measurements were inaccurate in the lower atmosphere because of local air pollution. So despite the fact that the ozonesonde data should have provided a comprehensive record of measurements over a two-decade period, many of the thousands of measurements received could not be used to determine once and for all whether or not ozone was increasing or decreasing in the troposphere or stratosphere.

In her research, Logan examined a set of data that contained more than 15,000 ozonesondes launched from the mid-1960s through 1982 (see Figure 17). In a 1985 paper published in the *Journal of Geophysical Research*, she concluded with evidence that ozone had in-

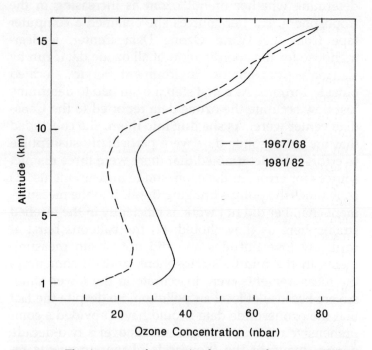

Figure 17. The increase of ozone in the troposphere over Hohen-peissenberg, West Germany, between 1967–1968 and 1981–1982. The data were derived from ozonesonde launches during this time. The concentrations are shown in nanobars (nbar).

creased throughout the troposphere over Europe at a
rate of 1 to 2 percent per year. Although most North
American stations showed a comparable increase, the
uncertainty of the measurement technique and the generally smaller data base provided a less certain foundation in support of the increase. Logan also analyzed
ozone measurements at certain ground stations, and
these also showed an increase in both Europe and
North America. But because these stations were located
in regions that might be readily influenced by local or
urban pollution, trend information using such data
would always be subject to the argument that an increase would not be truly representative of "global" or
background conditions.

For a more convincing set of measurements, scientists turned to the place where the first indications of
the carbon dioxide trend had been discovered: Mauna
Loa in Hawaii. As a result of the important information
obtained at Mauna Loa, the National Oceanographic
and Atmospheric Administration (NOAA) established a
monitoring network of four sites in the very remote
regions of the world (Mauna Loa; Barrow, Alaska;
American Samoa; and the South Pole) to keep track of
the long-term behavior of CO_2 and other important
trace gases that might alter the climate (see Figure 18).
The NOAA called this program Geophysical Monitoring
for Climatic Change (GMCC), and by 1973, the measurement program at the GMCC stations had been expanded to include the measurement of tropospheric
ozone. The measurements at these stations, including
Mauna Loa, are free of many of the problems Logan had
confronted in analyzing the ozonesonde data. These
stations are far removed from any sources of local pollu-

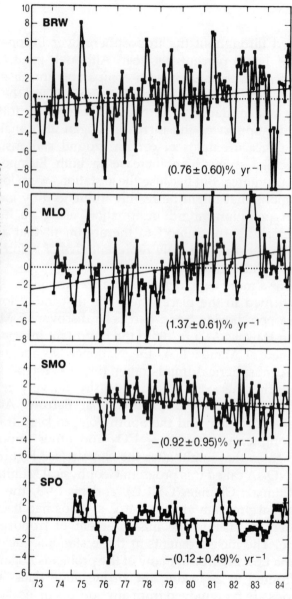

tion, and the instrumentation at each station is not only identical but has been improved so that it is not subject to the same types of interference as the instrumentation supplying the data for the ozonesonde measurements.

After more than a decade of measurements and careful analyses by research scientist Sam Oltmans and co-workers at NOAA, who take ozone measurements at the GMCC stations, the data obtained from the two northern hemisphere stations, Mauna Loa and Barrow, showed a significant increase in ozone on the order of 1 percent per year between 1973 and 1984.

In the southern hemisphere, neither the station at American Samoa, at 14 degrees south, nor the station at the South Pole, showed an increasing trend. In fact, both southern hemisphere stations exhibited a downward trend of 0.5 percent per year during this same time period. Why would the trend be downward in the southern hemisphere and upward in the northern

Figure 18. The annual trend at four surface ozone-monitoring stations. These stations are part of the monitoring network of the National Oceanographic and Atmospheric Administration's Geophysical Monitoring for Climatic Change (GMCC) program: BRW, Barrow, Alaska; MLO, Mauna Loa, Hawaii; SMO, American Samoa; and SPO, South Pole. The data indicate that surface ozone has increased significantly in the northern hemisphere, but not in the southern hemisphere. Note also how much variability is present in the monthly averages shown for these stations. Compare these measurements to the relatively smooth increase observed for carbon dioxide at Mauna Loa in Figure 1. (From S. J. Oltmans and W. D. Komhyr, "Surface Ozone Distributions and Variations from 1973–1984 Measurements at the NOAA Geophysical Monitoring for Climatic Change Baseline Observations," *Journal of Geophysical Research*, 91, 1986, pp. 5229–5236. Copyright American Geophysical Union. Reprinted with permission.)

hemisphere? Scientists speculate that the observed trend of ozone in the stratosphere over Antarctica might have some influence on the trend observed at the surface. But such an explanation is still speculative. Research continues today to answer such questions.

In another piece of research, Andreas Volz and Dieter Kley, scientists at a research institute in northern Germany, reconstructed Soret's instrument which had been used in the "Paris studies" between 1876 and 1910 (discussed in Chapter 1). Their research consisted of two important components: First, they carefully reconstructed the instrument used at the Montsouris Observatory on the outskirts of Paris and calibrated it using modern calibration techniques (see Figure 19). In this manner, they were able to compare the measurements reported in the records of the "Paris studies" with an accurate concentration that would be meaningful by today's standards. The second component of their research involved going back to the meteorological records between 1876 and 1910 to screen out the measurements when the wind was coming from Paris, thus making sure that they were measuring air that most likely did not contain any other pollutants which might interfere with the ozone in the air.

Volz and Kley discovered that air coming from Paris generally contained about 20 percent less ozone than air that had recently been over rural environs. They surmised that the lower concentrations from the city were the result of sulfur dioxide interference. Such a conclusion would be consistent with the fact that coal was used for home heating; it was likely that the coal used at the end of the 19th century was relatively high in sulfur content. Furthermore, they noted that the difference be-

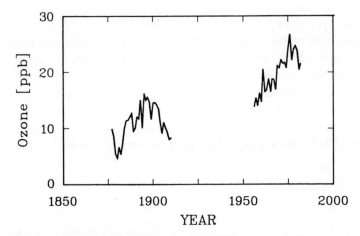

Figure 19. The annual mean ozone concentrations at Montsouris Observatory outside of Paris (1876–1910) and Arkona, East Germany (1956–1984). The average ozone concentration at the beginning of the 20th century near Paris was less than 10 parts per billion whereas in 1985 typical ground-level concentrations in central Europe were approaching 30 parts per billion, implying an increase of about 200 percent during the century. Scientists need to determine whether such an increase has occurred worldwide, or if this dramatic increase is confined to the lower atmosphere of central Europe. (After A. Volz and D. Kley, "Ozone Measurements in the 19th Century: An Evaluation of the Montsouris Series," *Nature*, 1988, *332*:240–242. Copyright MacMillan Journals. Reprinted with permission.)

tween the measurements of ozone of urban origin and ozone originating from other regimes was greatest during the winter, when coal usage was highest.

Volz and Kley concluded in their study that ozone concentrations in the background atmosphere outside Paris averaged 9 parts per billion (0.009 ppm) around the turn of the century. They claimed this estimate was accurate to 20 percent. Thus, in the 100 years since the Paris studies, it is likely that the content of ozone in the troposphere near the surface has at least doubled and, very likely, tripled. The implications of such an increase for our climate, for our environment, and for human health is the focus of the next several chapters.

CHAPTER FIVE

Hot Air

One summer day not long ago the Newport News (Virginia) *Daily Press* published an article about a weather forecaster who was predicting a major hurricane for the Hampton Roads, Virginia, area during the third week of September. The forecaster, according to the newspaper, had based his warning on the premise that the sunspot cycle was exactly right at that particular time to create conditions which would ensure that that particular stretch of the east coast would experience its worst hurricane since 1933. Years before, this same forecaster had predicted a severe hurricane season based on something he called the "banana blow-down index."

Sunspots and bananas. With all our technology, forecasting the weather is still not as exact a science as we would like it to be. The weather, after all, is so *unpredictable*, so changeable. Is it any wonder that men and women throughout time have sought for some simple explanation, some definite pattern, upon which they could accurately and forthrightly predict the future?

Sunspots at one time were thought to hold the key.

In the 1940s several astronomers were able to use advanced telescopes and other devices to study the sun. They discovered what appeared as dark spots on the sun's surface, which, upon examination, turned out to be solar "storms." With further study over the following years, scientists determined that there appeared to be a cycle of such storms occurring and of the ebb and flow of storm activity on the sun's surface, which also appeared to affect earth. For example, during very active periods of sunspot activity, radio communication here is often disrupted, and the auroras (the so-called northern and southern lights) are more active. This particular cause–effect relationship led other fields of science to look for more. Thus some meteorologists began to correlate weather, which also seems to be organized cyclically, with sunspots. For a time during the 1950s and 1960s, weather forecasters were predicting hot summers or cold winters based on where we were in relation to the 11-year sunspot cycle. The theory was that since during periods of high sunspot activity summers in the northern hemisphere tended to be warmer, such warm summers would also produce more hurricanes. Conversely, when sunspot activity was at its low point in the cycle northern hemisphere summers would be cooler, and there would be fewer hurricanes. This theory seemed to hold true for a decade or so, during the late 1950s, but as more data were compiled and newer technology gave meteorologists more tools, the sunspot theory as a tool for weather forecasting proved unreliable.

What about the banana blow-down index? Back in the 1960s a meteorologist discovered that there was a high correlation between the number of bananas lost to high winds in Panama and the number of hurricanes generated in the Atlantic Ocean the following autumn. This may sound farfetched, but there is a scientific basis for the banana blow-down index. The presence of high winds prevailing in Panama during the summer is linked to the same meteorological phenomena that produce favorable conditions for the formation of hurricanes in the Atlantic during hurricane season.

Both the banana blow-down and the sunspot theory are prime examples of how interrelated the earth's atmosphere is, of how conditions halfway around the world and even out of this world can affect the nature of things in other faraway places with strange-sounding names. Like sunspots, for a short time the banana blow-down theory was thought to be fairly accurate, but over the long haul it, too, proved a failure as a dependable forecast tool. Weather is just too dynamic, with too many variables, to be summed up by one forecasting technique. And by the way, no hurricane even came close to Hampton Roads, Virginia, that particular September, or for the entire hurricane season, either. So much for the sunspot–hurricane connection.

More recently, in 1988, a scientist testified before a congressional subcommittee that global warming due to the increase of carbon dioxide was already here, already occurring, and was not a science-fiction scenario for the distant future. This testimony was based on a thorough analysis of existing temperature records dating back to the 19th century, and on the results of some of the most advanced computer tools available anywhere in the world. But are these enough to make such a testimony reliable? The public is already aware that meteorologists and weather forecasters have the most advanced satellite and computer data available to them, yet we are still disappointed by inaccurate forecasts. Is this prediction about global warming true or, like sunspots and the banana blow-down theory, will such a prediction prove once again that nature refuses to be second-guessed? Perhaps only time will tell.

WEATHER BY THE NUMBERS

So if you can't forecast the weather by counting bananas, or by counting sunspots, what can you use? Computers. But to understand how computers help forecast weather, we need to go back in time to when the study of weather was a fledgling science. Back in 1812, the French mathematician Pierre Simon de Laplace wrote that complete knowledge of the masses, positions, and velocities of all particles at any single instant would enable precise calculation of all past and future events. This was and is the basic philosophical premise of weather forecasting, that the physical laws governing the atmosphere can be used to determine its future state. During Laplace's time scientists were already familiar with the classical laws of mechanics as described in 1687 by Sir Isaac Newton in his classic treatise, *Principia Mathematica*. But what Laplace and his contemporaries didn't know at the time were the fundamental laws of thermodynamics, which are crucial for predicting the motions and temperatures of fluids such as air and water.

This void in knowledge was filled by the middle of the 19th century, when the German scientist Rudolf Clausius identified the conservation of energy as the

first law of thermodynamics and then went on to formulate the second law of thermodynamics. In the absence
of external constraints, the net flow of heat between two
bodies is from the warmer to the cooler one. Which is
why when you place a hot cup of coffee on the table it
will eventually cool. By the end of the 19th century the
world had recognized the fundamental laws of classical
physics, and the goal of accurately predicting outcomes
by numerical calculations was closer to reality. In applying the general laws of physics to the atmosphere, the
Norwegian meteorologist Wilhelm Bjerknes wrote in
1904:

> If it is true, as every scientist believes, that subse-
> quent atmospheric states develop from the preceding
> ones according to physical law, then it is apparent that
> the necessary and sufficient conditions for the rational
> solution of forecasting problems are the following:
> 1. A sufficiently accurate knowledge of the state of
>   the atmosphere at the initial time.
> 2. A sufficiently accurate knowledge of the laws ac-
>   cording to which one state of the atmosphere
>   develops from another.

It was up to the British scientist Lewis F. Richardson to take the significant step of applying this meth-
odology to weather forecasting. He began his work on
numerical weather forecasting in 1913 while working as
superintendent of the Eskdalemuir Geophysical Observatory. When World War I broke out Richardson, a
Quaker, filed for conscientious objector status, which
allowed him to continue his work while tending to ambulance runs as a member of the Friends' Ambulance
Unit in France. The fruition of his work was the monumental *Weather Prediction by Numerical Process,* published

in 1922. But the Great War had taken its tool on the scientist. As O. M. Ashford wrote in Richardson's obituary in 1953:

> As might be expected in a man of Quaker upbring-
> ing, Dr. Richardson often wondered how his abilities
> could best be used to serve mankind, and he gradually
> turned away from meteorology to pioneer a completely
> new subject: a mathematical and psychological analysis
> of the causes of war. How rare it must be for a scientist
> honoured with a D.Sc. and F.R.S. [Fellow of the Royal
> Society] (in 1926), to start studying and take a B.Sc. in
> psychology (in 1929). His new researches consumed
> more and more of his time and one of the main reasons
> for his early retirement [from the Geophysical Obser-
> vatory] was to enable him to devote all his energies to
> this task. From then on meteorology was regarded as
> temptation to be resisted, so that his thoughts could be
> concentrated on his service to world peace. *Generalized
> Foreign Politics*, published in 1939, showed how this ap-
> parently abstruse mathematical work could produce
> practical value.

But even more disheartening to Richardson was that despite all the work that went into his calculations, his forecasted state of the atmosphere proved to be a complete bust. He knew that the key to his equations was a knowledge of the meteorological variables at several levels of the atmosphere. The atmosphere, as we've discussed previously, occurs in "layers," the layer closest to Earth being called the troposphere. But the troposphere itself is not uniform: with every increment of space above the surface, the air can be different. To formulate his calculations Richardson chose five layers of the atmosphere to study, each spaced two kilometers (a bit more than a mile) apart. He chose to provide forecasts for two points over central Europe, but to do so he needed to know the winds, temperatures, and pressure at all five vertical levels at each of 18 different, equally spaced locations. Richardson worked up a grid, positioning each of the 18 locations three degrees of longitude (about 300 kilometers) from each other east to west, and separating them by about 200 kilometers north to south. Such spacing would, he assumed, give him enough data vertically (via the five different levels) and horizontally (via the grid covering all four directions) to work with. After he had analyzed and calculated, he came up with a prediction of a six-hour pressure drop of 145 millibars over central Europe. Such a drop had never before occurred, or at least had never been observed, in nature. To give an idea of how far out such a prediction was, the most intense hurricane of this century, Hurricane Gilbert in the summer of 1988, with sustained winds of over 200 miles per hour, produced a central pressure that was only 120 millibars lower than the pressure outside the hurricane.

The blown forecast was quite a setback for the field of numerical weather prediction. For three decades after Richardson's work, until the introduction of electronic computing machines, the field remained stagnant. After World War II a new era began, thanks to the development of the electronic numerical integrator and computer (ENIAC). The ENIAC had originally been developed for military purposes during the war at its home base, the Ballistic Research Laboratory of the Aberdeen Proving Ground in Aberdeen, Maryland. In 1946, John von Neumann, a professor of mathematics at MIT who had been a consultant at the Ballistic Research Laboratory, submitted a proposal to the U.S. Navy to create a mete-

orology group to use the ENIAC. The group could then
duplicate what Richardson had hoped to do more than
three decades earlier, but with the help of the new-
fangled electronic computing machine.

But computers aren't miracle makers. Calculations,
whether they are done by hand or by computer, are still
calculations; the outcome is determined by the input.
But what made von Neumann's proposal a success was
that during the three decades since Richardson's failure,
scientists had got some important insights into the basic
physics that govern meteorological process equations.
Carl-Gustav Rossby, for example, a Swedish meteorologist generally regarded as the "father of modern
meteorology," demonstrated in 1939 that certain assumptions could be made about the movement of air in
the atmosphere that allowed for much simpler equations than those Richardson had worked with. In his
classic paper, Rossby showed that there were two large-
scale forces in the atmosphere: one was created by the
rotation of earth, called the *Coriolis force;* the other was
the effect of atmospheric pressure, which, according to
Rossby, was not necessarily the same at every point.
The latter effect is called the *pressure-gradient force.* So
what Rossby surmised was that these two forces were
trying to balance each other and that that was all you
had to know to calculate the fundamental equation that
Richardson had so diligently worked on. What Rossby's
theory meant was that many of the processes Richardson had included in his equations could be neglected,
such as the small-scale energetics of intense heating of
the earth's surface responsible for summertime thun-
derstorms. In fact, the inclusion of such processes could

lead to erroneous results for the larger scaled calculations that Richardson had attempted.

The U.S. Navy funded von Neumann's proposal in
1946. Von Neumann sought some kind of commitment
from the meteorological community, including Rossby,
who at that time was professor of meteorology at the
University of Chicago, and Harry Wexler, director of the
National Weather Service. The result was the formation
of the Institute for Advanced Study at Princeton, New
Jersey. It was here that the first "successful" experiments in numerical weather prediction were achieved
in the early 1950s. One of the experiments, conducted
by Norman Phillips, Jule Charney, and Joseph Smagorinsky, produced a numerical model that accurately describes the physics necessary to forecast the weather,
just as Rossby had postulated in 1939.

Smagorinsky later became the first director of the
Geophysical Fluid Dynamics Laboratory, an autono-
mous facility of the National Oceanic and Atmospheric
Administration (NOAA), which develops computer
models for the purpose of forecasting weather and simulating climate. By 1955 the National Weather Service
had begun on a trial basis to issue forecasts based on
numerical predictions. By 1960 these forecasts had been
made routine, coming out of the National Meteorological Center in Washington, D.C., for use by any
and all meteorologists throughout the country. In the
following decades the predictions continued to improve
as the speed and capacity of computers grew, and as
more and more data, especially from satellites, came to
be utilized in the equations—the same basic equations
that back in 1904 Bjerknes had insisted must be well

known in order for the weather to be accurately forecasted.

CLOUDY ISSUES

Still, despite the improvement, weather forecasting is not an exact science. Everyone probably has a favorite tale of a weather forecast that didn't come true. The family picnic that was planned after the weatherman called for a beautiful day is ruined when morning brings clouds, fog, or snow. Or the skier waxing his skis after hearing on the radio a forecast of six to eight inches wakes the next morning to sunshine and clear skies and not a single flake of the white stuff. Such occurrences are quite common, even in this age of satellites and high-speed computers, because the weather on earth for the most part is changeable, and only certain aspects of it are predictable. So if the weather is that difficult to predict, you can imagine how difficult it must be to forecast the earth's *climate*. There is a difference between climate and weather. Weather is the general condition of the atmosphere (e.g., clear, cloudy, rainy, or foggy) at a particular time and place. Climate is the *prevailing* weather at a particular place over a long period of time.

Before climate can be predicted, it must first be defined. To obtain such a definition requires the analysis of a huge number of historical records—the more data and the further back in time it goes, the better the chance of an accurate definition of climate. To help forecast climate changes, scientists often work with models. The initial goal of climate models is to simulate the present climate patterns as determined from the available meteorological observations, including data that go back in history. But even given the technology and computers that make such model simulations possible, the task of forecasting climate is very difficult because the earth-climate system is so very complex.

As an example of this complexity, try to answer the following question: "Does the presence of clouds heat or cool the earth?" As discussed earlier, we know that water vapor in the clouds traps some of the radiant heat as it leaves the earth's surface and reradiates it back to earth. This is what farmers pray for to prevent an early frost, for they know frost is unlikely on a cloudy night. But we also know that the presence of clouds blocks out the sun's rays and therefore prevents heat and light from reaching the earth's surface. So on a global basis, which effect is more pronounced?

Since 1984, in an attempt to answer that question, scientists have been monitoring a sophisticated set of three satellite instruments as part of the Earth Radiation Budget Experiment (ERBE). Because of the complexity of the problem, and in order to arrive at a reliable and credible answer, scientists had to solve the problem of scale. They could look at one cloud, over one part of the earth, and come up with a lot of data. But that's not enough. They must be able to look at more than the detailed physics of one particular cloud, or even several clouds, to understand the whole picture: how clouds interact with the earth's climate. Looking at one small part of the earth, no matter how many data you accumulate, is like taking the temperature, humidity, and barometric pressure in your back yard and applying those readings to your entire state. What scientists had to do was to look at many clouds over many different parts of the earth. Satellites have made such observations possible. Satellites can measure the properties of nearly all the clouds around the world. The task is formidable: scientists must examine the different properties of many different types of clouds that are constantly in motion, sometimes over land, sometimes over water, or over mountains. Why is it that certain types of clouds seem to prefer oceans rather than land? And why do some clouds appear on one side of a mountain range but never on the other?

As a starting point to understanding the complexity of the earth-climate dynamics, we can view the earth itself as a "black body." In physics, a black body is an object that reemits all the energy it receives. It is not necessarily of the color black. Satellite measurements show that 327 watts of energy intercept every square meter of the top of the atmosphere. If we could convert this energy into electricity, we'd have enough to light more than five 60-watt light bulbs for every square meter of matter upon which the sunlight falls. Based upon the amount of radiation emitted by the sun, and the fact that there is a direct relationship between the quantity of energy that hits the earth and how hot the earth should become, the earth's black body temperature should be about 0 degrees Fahrenheit. In reality, however, the earth's average temperature is nearly 60 degrees Fahrenheit. Why? Because of the greenhouse gases in the atmosphere, primarily carbon dioxide and water vapor.

That's the easy part. What's more difficult to answer is just *how* the heat stored in the atmosphere and on the land and ocean surfaces is retained and redistributed throughout the planet. Scientists call this dynamic feature of heat transfer around the globe by winds, oceans, clouds, and other atmospheric processes the *general circulation of the atmosphere*. The temperature patterns and precipitation distribution measured as a result of the general circulation are known as the earth's climate. To study the global climate and its potential variability, scientists have traditionally used extremely complex computer models that simulate the general circulation of the atmosphere.

Because the low latitudes of the earth (i.e., the area near the equator) receive more heat than the high latitudes (near the poles), and because the nature of heat is to expand and move, heat is transported from the tropics to the mid and high latitudes. Some of this heat is moved by winds and some by ocean currents, and some gets stored in the atmosphere in the form of latent heat of evaporation, or just plain latent heat. The term *latent heat* refers to the energy that has to be used to convert liquid water to gas, or water vapor. We know that if we heat (not boil) a pan full of water on a stove, it will evaporate faster than if it is allowed to sit at room temperature. We also know if we hang wet clothes outside in the summertime they will dry faster than in the winter, when temperatures are colder. The energy in both cases is supplied by heat, by the stove in the first case, by the sun in the latter case. But what happens to this energy? It's stored in the atmosphere as latent heat. Eventually the water stored as a gas in the atmosphere will condense to a liquid again, and the energy will be returned to the atmosphere.

In the atmosphere, a large proportion of the incoming sun's energy is used to evaporate water, primarily in the tropical oceans. In an article appearing in *Physics*

Today, Vaerabhadram Ramanathan, an expert on global climate from the University of Chicago, and two of his colleagues—Bruce Barkstrom and Ed Harrison, both from NASA Langley Research Center, which oversees the operation of the ERBE data-processing—tried to quantify this proportion of the sun's energy. By analyzing temperature, humidity, and wind data around the globe, they estimated the quantity to be about 90 watts per square meter, or nearly 30 percent of the sun's energy. Once this latent heat is stored within the atmosphere, it can be transported (primarily to higher latitudes) by prevailing, large-scale winds or it can be transported vertically to higher levels of the atmosphere, where it forms clouds and subsequent storms, which then release the energy back to the atmosphere. As anyone who has climbed mountains knows, the air temperature changes as one rises in altitude, usually becoming about three Fahrenheit degrees cooler with each 1,000 feet of elevation up to the altitude of the lower stratosphere. The storm process caused by vertical heat transport changes the rate at which the temperature decreases as the altitude increases. The formation of clouds allows less of the sun's radiation to reach the earth's surface. So it's easy to see why the answer to such a simple question as whether the presence of clouds heats or cools the earth is not so simple, even though scientists have spent years analyzing monumental amounts of data from three satellites just to get a handle on it.

So far, analysis of ERBE data has revealed that the presence of clouds acts as a cooling force on the earth's climate system. Before ERBE, scientists weren't sure if the earth was warmer or cooler because of the presence of clouds. The ERBE results show that clouds absorb, on a global basis, an average of about 30 watts of energy per square meter of the radiation that comes back from the earth's surface after being warmed by the sun's energy. The amount of energy reflected by the presence of clouds, on a global basis, is about 47 watts per square meter. Thus there is a net cooling of the earth's atmosphere of about 17 watts per square meter. What that means is that without clouds, the planet would be warmer. The magnitude of the warming would depend on exactly how the energy balance was converted to temperature. For most global-climate models (GCMs), the temperature increase would be on the order of 10 to 15 degrees Fahrenheit. For the first time, such computed results from GCMs can be compared to a truly global measurement of the earth-climate system to see how well they duplicate cloud processes and what effect they have on perturbations in the earth-climate system. According to such calculations, doubling the amount of carbon dioxide in the atmosphere would alter the heat balance by only four watts per square meter, an effect of only one-fourth that of the presence of clouds.

There are many parameters within the GCMs that can be checked by observations. For the most part, GCMs do reasonable jobs of simulating temperature, precipitation, and winds. But none of them simulates these observed meteorological variables perfectly, and scientists are constantly trying to improve GCMs by "tuning" them. *Tuning* simply means adjusting the numerical values of certain parameters in the model to make them agree as closely as possible with the observational data. If too much tuning takes place, if the model contains too many trouble variables and the observational data are of limited size, then the results, though they will "fit" precisely, will not be credible. That's why projects like ERBE, where scientists are developing new observational data in relation to which GCM results can be analyzed and compared, are so valuable. Given all that, the problem of modeling the earth's climate is still difficult. Quantifying the impact of increasing carbon dioxide concentrations on the earth's climate is not a trivial task, and the results from such theoretical calculations must be taken with some skepticism, regardless of how sophisticated the computer models are.

TELLING IT LIKE IT IS

On June 23, 1988, Dr. James E. Hansen testified before Congress that he was 99 percent sure that the relatively high temperatures experienced in the United States during the 1980s were a sign that the greenhouse effect resulting from higher levels of carbon dioxide was already upon us. Hansen's credentials are impressive. At the time of his testimony he was director of NASA's Goddard Institute for Space Science (GISS) in New York City. GISS is a relatively small institute by government standards, comprised of a few dozen scientists and computer programmers located in one building close to the campus of Columbia University. GISS is not to be confused with the Goddard Space Flight Center, which is a much larger institution with thousands of scientists and technicians, located in Greenbelt, Maryland.

GISS is one of the few institutes in the world that

has developed a general circulation model. When given the assignment of simulating the effect of increasing carbon dioxide on the earth's climate, the GISS general circulation model has produced results that are consistent with the results of other general circulation models, namely, that temperature will rise as CO_2 is increased. The detailed results from the GISS model were published in 1988 in the *Journal of Geophysical Research*.

General circulation models are extremely complex and produce an overwhelming number of data. The GISS model is set up to solve three important primary equations: (1) to describe the momentum of the winds; (2) to describe the heat transfer which occurs when the sun's heat is modified by the absorption of the different types of surfaces on the planet and is then redistributed by clouds and winds; and (3) to describe the subsequent transport of water vapor, which is complicated by the process that produces clouds and precipitation. These basic equations are solved at nine different vertical levels in the atmosphere, on a grid that covers 8 degrees latitude by 10 degrees longitude. In other words, the model can compute those three equations at nine different altitudes above the earth. The equations also take into account such factors as the distribution of the major water features of the world, the different reflectivity of the earth's surface due to snow or ice, and the presence of mountains. Such knowledge is not cheap. These models require large, powerful computers and the commitment of massive amounts of computer resources (i.e., time and programmers) to run them.

So once a general circulation model is run in a computer, what does the team of scientists look for to check how well it runs? In principle, the equations are not that difficult to understand; they describe the evolution of temperature fields, precipitation patterns, and wind distribution throughout the globe. To come up with such patterns, the model uses only the sun as its input of energy, and the resultant fields evolve as a function of the earth's distribution of land and sea, the position of mountains, and the fact that the energy driving the weather on this planet is not uniformly distributed over its surface. Add to this mixture the fact that this sphere we call earth rotates ever 24 hours and the energy distributed from pole to pole goes through a yearly cycle that determines what we call seasons, and you have a remarkable number of variables. The general circulation model converts all this into measurable meteorological quantities, what scientists call *general circulation statistics*.

These statistics relate to both spatial and temporal computations, that is, to the computation of temperature and precipitation through space and time. For example, if the model describes the type of climate that would put a desert in northern Africa, how big is the desert? Spatial relationships are difficult to interpret because, as mentioned earlier, the model is constrained to a certain resolution. In the case of the GISS general circulation model, their 8-degree latitude by 10-degree longitude translates into boxes approximately 500 miles by 600 miles. Unfortunately, physical climatic features, such as deserts and tropical rain forests, are seldom exactly rectangular and do not in general fit into areas on the planet that are bounded exactly by the locations of the boxes that constrain the model output. In other words, part of the box within the general circulation model may contain an observed area that is in reality one-fifth desert and four-fifths savannah. Thus it is not a straightforward task to compare the climate of that box with the meteorological records directly observable over that region since there may be a high degree of variability within the box itself. This sounds formidable, but all is not lost.

Once the means by which the spatial interpretation of the computed data that can be analyzed is determined, the seasonal and long-term temperature and precipitation information over a particular region can be examined and compared with existing data. That is, scientists can compare what the computer says with what the actual observational data say. For example, if we consider the area of the United States known as the Great Plains, and run a general circulation model of it, we can ask certain questions. In the grid box of the general circulation model that defines the region, we can compare the temperature in the box with the average temperature actually recorded over a 100-year period. Does the average temperature look reasonable? Does the difference between mean summer and winter temperatures look reasonable? Is the average amount of rainfall and snowfall computed by the model consistent with observations, and does it rain more in certain months than in others? Once these statistics are analyzed, more complicated interpretations can be examined.

If the general circulation model were allowed to run for 100 years, we could examine its output to determine how many times during that period the model said drought conditions would exist on the Great Plains. So it is important that the models be able not only to predict average conditions, but also to predict how frequently and how severely temperature and precipitation differ significantly from the average. A general circulation model might, for example, do a perfect job of simulating the average climate over Kansas and at the same time fall far short of describing how many extremely dry and hot summers may occur over a given period of time. And even if the model does a perfect job of creating the proper statistics for Goodland, Kansas, it may fall short of simulating the climate over Paris, France. Thus the general circulation model team is faced with the extremely tedious task of comparing an immense number of data from all over the world with the model output. If differences between the model output and the observations exist, the team must try to determine the reason for such differences and then see what approximations they made in their equations that must be modified.

Scientists working on the GISS general circulation model began by establishing a baseline. They did this by feeding data from meteorological records over the past 100 years into the computers. The concentrations of carbon dioxide (315 ppm) and methane (1.5 ppm) were fixed at 1958 levels, as were those of nitrous oxide and chlorofluorocarbons. After a computer run of 100 years (a simulation of 100 years on the computer), the average global temperature during that period did produce an annual variability of about one degree Fahrenheit. In other words, the average temperature for one year of the model run, when averaged over the entire planet, would be as much as one degree warmer than the temperature of another year. This tendency is referred to as *interannual variability,* and it can be quantified. It's an important variable, and when running a model on a general circulation model, the amount of variability must be simulated if the results are to be credible. With the GISS general circulation model, the agreement between the observed interannual variability and the model's computed interannual variability is considered quite good.

Once scientists have a baseline (i.e., the baseline is the results for the 100-year run, with fixed carbon dioxide and other trace gas concentrations), the model is run again for 100 years, but this time with the trace gases allowed to increase. But how much should they increase? Here scientists have to make some assumptions. One way they can prescribe trace gas concentrations is to take the increases actually observed during the 1980s and feed them into the model, so that from 1958 through the 1980s, the model is using actual concentrations of these trace gases in its computations. To predict the next 100 years, however, requires more speculation. For the GISS general circulation model, scientists set the rate of increase for these gases at the same rate as occurred between 1958 and the early 1980s. In other words, by taking the current rate of increase and projecting it to continue to the year 2060, scientists hoped that the model would present an accurate forecast. As the concentrations of these trace gases increase, more of the radiation emitted from the earth's surface will be absorbed by them and then reradiated back to earth. After 100 years, according to the model, the average global temperature will be approximately seven degrees warmer. That is, in the year 2060, the average temperature of the earth will have risen seven degrees Fahrenheit from that of 1960.

Most of the general circulation models in use produce fairly consistent results. The GISS results are consistent with those of other models used, but it must be remembered that such results are still merely computer simulations and therefore, as scientific tools, are imperfect. As mentioned earlier, the science of atmospheric chemistry is one of the few sciences impossible to study in a controlled laboratory setting. The earth is the only laboratory, and it is impossible to control. The kind of modeling just described helps produce our weather forecasts, and everyone knows how accurate they are. Despite this imperfection, computer modeling has improved the science of forecasting considerably, although such forecasts remain part science, part art.

HOW HOT IS IT?

When Hansen testified during the congressional hearing in 1988, the comment that seemed to set apart his testimony from that of other experts was that he thought the greenhouse effect was already upon us. Such a bold statement had never been made before by such a reputable scientist. Hansen based his belief on one primary fact: temperature records show clearly that four of the hottest years in the last century occurred in the 1980s. His conclusions, however, immediately drew fire from other experts in the climate community. Was Hansen right? Or was he jumping to conclusions?

Hansen based his statement on an analysis of temperature data from more than 1,000 weather stations covering approximately 80 percent of the earth's surface (see Figure 20). These temperature records show that temperatures had indeed increased over the past century, on average from .9 to 1.2 degrees Fahrenheit. True, the early part of the 1980s was particularly hot, period

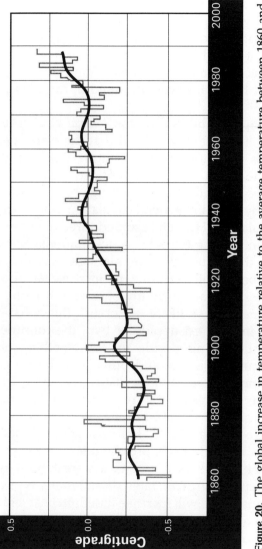

Figure 20. The global increase in temperature relative to the average temperature between 1860 and 1988. The solid line is a five-year running average. The individual yearly averages are shown by the light bars. The warmest four years during this period have occurred during the 1980s. (Figure courtesy of the National Oceanographic and Atmospheric Administration.)

in the 1930s during the famous "dust bowl" years of the Depression was equally warm. Despite the hot 1930s, the temperatures cooled again during the 1960s and 1970s, and it was this trend that was thrown back into Hansen's face after his statement. How do we know there won't be another cooling trend in the 1990s, just as before?

In a paper published in the *Journal of Geophysical Research* in 1987, Hansen argued that the warm period in the 1980s was more global in nature than the warm period of the 1930s, which was confined to the northern midlatitudes. Figure 21 shows the global temperature computed by the GISS general circulation model. In this model simulation, Hansen assumed three different scenarios for increasing trace-gas concentrations. Scenario A assumes that carbon dioxide, methane, and the chlorofluorocarbons will continue to increase at approximately the same rate currently observed. It is the scenario deemed most likely to occur. But what will happen if, in the next decade or two, the countries of the world unite and decide to do something about dumping carbon dioxide and other greenhouse gases into the atmosphere? If a realistic, concerted global effort to curtail emissions is realized, then the model computes a smaller global warming effect. Nonetheless, even with our best efforts, temperatures will still increase. The amount of this projected increase is shown in Figure 21 by curves B and C. Whereas Scenario A assumes that the growth rates of emissions typical of the 1970s and 1980s will increase indefinitely at a rate of 1.5 percent annually, Scenario B assumes that trace gas emissions will increase by the same amount in the future as they did during the 1970s and 1980s. Thus, although the increase of emissions in Scenario B is 1.5

percent now, it is less than that in years to come. Scenario C represents a more dramatic curtailment of emissions. It represents elimination of *all* chlorofluorocarbon emissions by the year 2000 and a reduction of carbon dioxide and other trace gas emissions to such a level that its annual growth rates are zero by the year 2000.

The agreement between the model calculations and the temperature analysis between 1960 and 1985 is remarkably good. These calculations take into account the observed increase in concentrations of carbon dioxide, methane (see Figures 22 and 23), and the chlorofluorocarbons. However, it is easy to see why many of Hansen's colleagues think his results cannot be extrapolated "with 99 percent accuracy" to the year 2060. In Figure 21, the heavy solid line represents the temperature observation from all over the world. Although the model results through 1985 (the last year of temperature data plotted on this graph) are relatively good, very few of Hansen's contemporaries believe that it is proof that temperatures will rise at the rate predicted by the model. In this case, the only proof is time. But can we afford to wait?

One of the most intriguing parts of Hansen's study is his discussion of temperature changes in certain cities of the United States in years to come. Limiting the discussion to Scenario A (i.e., the most likely case, in which carbon dioxide, methane, and chlorofluorocarbon concentrations continue to rise at the same rate as is currently observed) and taking the city of Washington, D.C., as an example, Hansen reported that the temperature records for the three decades between 1950 and 1980 show that a typical summer had an average of 6 days on which the temperature rose to above 95 degrees. The general circulation model predicts that by

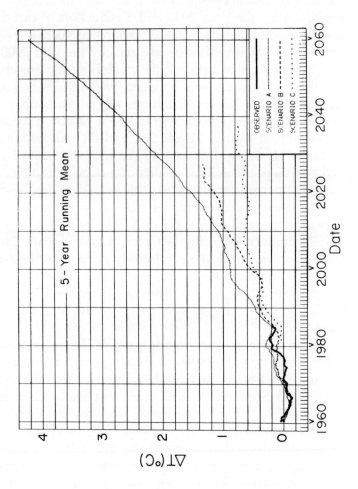

Figure 21. The observed average global temperature between 1960 and 1985 is shown by the solid line. These observations are compared with the calculated global temperature by the Goddard Institute for Space Studies general circulation model under three sets of assumptions regarding the rate of release of trace gases to the atmosphere between 1985 and 2060. Scenario A assumes that the growth rates of trace gas emissions of the 1970s and 1980s will continue indefinitely; the assumed annual growth averages about 1.5 percent of current emissions, so the growth rate of the concentration of the radiative trace gases increases exponentially. Scenario B has decreasing trace-gas growth-rates, so that the annual increase of greenhouse climate-forcing remains approximately constant at the present level. Scenario C drastically reduces trace gas growth between 1990 and 2000, so that the greenhouse climate-forcing ceases to increase after 2000. (Figure from J. Hansen et al., "Global Climate Changes as Forecast by Goddard Institute for Space Studies Three-Dimensional Model," *Journal of Geophysical Research*, 93, pp. 9341–9364. Copyright American Geophysical Union. Reprinted with permission.)

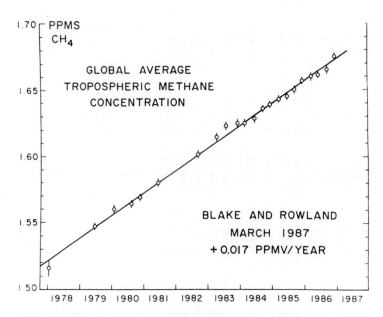

Figure 22. The increasing average global concentration of tropospheric methane between 1978 and 1987, from measurements obtained by Donald Blake and F. Sherwood Rowland of the University of California at Irvine. Methane is also a greenhouse gas, so that increases in its concentration contribute to global warming. Methane's rate of growth is 0.017 ppmv (parts per million, by volume) per year. (Figure from D. H. Ehhalt, "How Has the Atmospheric Concentration of CH_4 Changed?" in *The Changing Atmosphere*, F. S. Rowland and I. S. A. Isaksen, eds. Chichester, England: Wiley, 1988, pp. 25–32. Reprinted with permission.)

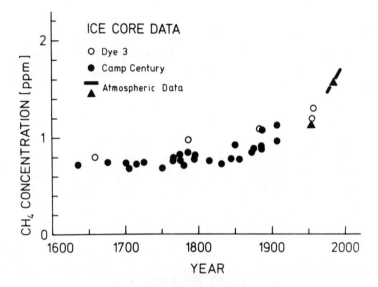

Figure 23. The increasing concentration of methane from the early 17th century through the 1980s. This trend has been determined from ice core samples (circles) from two different sites and atmospheric measurements. The preindustrial concentration was found to be less than 0.8 parts per million; the implication is that the natural concentration of methane has already more than doubled. (Figure from D. H. Ehhalt, "How Has the Atmospheric Concentration of CH_4 Changed?" in *The Changing Atmosphere*, F. S. Rowland and I. S. A. Isaksen, eds. Chichester, England: Wiley, 1988, pp. 25–32. Reprinted with permission.)

the decade 2000–2010 there will be 14 days, on average, each summer on which the maximum temperature will exceed 95 degrees Fahrenheit. By the decade of the 2050s, Washingtonians can expect 49 days with temperatures that high. And there won't be much relief at night. Climatology of the region shows that currently there are 9 nights when the temperature fails to fall below 75 degrees. By the 2000–2010 decade the model shows 22 such nights, and by the 2050s the computer predicts 60 such "sticky" nights during a typical Washington summer. Conversely, the winters will be milder if the future coincides with computer models. During the last three decades, for example, there have been on average 73 days per year when the temperature has fallen below freezing. By 2000–2010, there will only be 43 such days, and by the 2050s only 28 days. Such changes represent quite a departure from current climate.

HEAT ISLANDS

One factor in the dissension over the GISS study was the argument over the "urban-heat-island effect." Scientists reviewed Hansen's earlier study, published in 1987 in the *Journal of Geophysical Research*, and argued that the temperature data from which the 100-year trend had been derived had been strongly influenced by the urban-heat-island effect, a phenomenon caused when the local urban temperature is influenced by the presence of streets and buildings. The concrete and asphalt found in large urban centers absorb heat from the sun more efficiently than other types of natural surfaces. Scientists have documented many cases where the temperature of an urban area is up to several degrees warmer than that of the outlying countryside. But Hansen and his associates at GISS were aware of the heat island effect when they did their temperature trend analysis dating back to the middle of the 19th century, and they avoided using stations at sites with populations greater than 100,000. By comparing the 100-year temperature trend only at stations with populations over 100,000 to the remainder of the data available to them, the GISS scientists concluded that inclusion of these highly populated areas would create a bias of less than 0.4°F, or less than one-third of the trend they claimed to be present between 1880 and 1985.

Subsequently, however, some scientists have argued that no temperature records in areas of more than 1,000 population should be used for such an analysis. If we limited our study to only rural areas in the United States, for example, we would find a slight *cooling* between 1920 and 1985 over a large portion of the southeastern U.S. So even with the same data available to everyone interested in the problem of global warming, analyses of such data often reveal results which can be interpreted in completely different ways to support a particular hypothesis. This is why figuring out whether or not the earth is actually warming because of the greenhouse effect is such a challenging problem.

HOW REAL IS THE OCEAN? I'LL TELL YOU NO LIE

The general circulation model and its cousin, the present-day forecast model, are similar in function; they both crank through the basic equations that describe weather that can be observed. But general circulation models must be even more sophisticated, because they must include a simulated ocean, whose behavior becomes an integral part of how heat is transported around the globe. At the time Hansen published his results in the *Journal of Geophysical Research*, they were the first to describe output from a general circulation model that had any kind of realistic ocean and was also driven by realistically increasing greenhouse gases. Thus, at the time, his results provided the best guess of how climate might be responding then and how it might respond in the next few decades.

But other modelers disagreed with Hansen's results, including Stephen Schneider, a well-known climate modeler from the National Center for Atmospheric Research (NCAR) in Boulder, Colorado. In an interview appearing in the June 1989 issue of *Science*, Schneider stated that the GISS general circulation model had been coupling its atmosphere to a "pretty hokey ocean." Schneider wasn't singling out the GISS model as much as calling attention to the fact that most models suffered from the same shortcoming, and that all climate modelers should feel uncomfortable touting their results with as much confidence as Hansen had before Congress in 1988.

BUT WHAT ABOUT OZONE?

As complex and sophisticated as the general circulation models are, and as good a job as they do predicting the impact of increased carbon dioxide concentrations, none of the models yet includes the processes that describe the atmospheric photochemistry discussed in Chapter 3. So none of the general circulation models used for climate forecasting can accommodate a realistic scenario regarding tropospheric ozone. The GISS model, in its scenario for the future, assumes that ozone in the troposphere increases by only 25 percent while, during the same time, carbon dioxide and methane concentrations increase 100 percent.

Why Hansen and his colleagues at GISS assumed such a relatively slow growth rate for tropospheric ozone in their calculations is that the inclusion of tropospheric ozone in such models is a considerably more difficult procedure than the procedure for all other greenhouse gases. The rate of growth of tropospheric ozone, because it is not long-lived in the atmosphere, varies from one location to another. Such variability is not true for carbon dioxide, methane, or chlorofluorocarbons, which can be assumed to be well mixed throughout the troposphere because they are so long-lived. Furthermore, as we showed in the last chapter, the rate of growth of tropospheric ozone also appears to be different according to altitude. Theoretical considerations have shown that ozone will contribute more to the greenhouse effect if it is located in the middle and upper troposphere (e.g., between 20,000 and 40,000 feet) than if such an increase occurs near the earth's surface. Thus, for a variety of reasons, it was easier not to worry about the effect of a potential increase of tropospheric ozone on climate. Besides, no scientific study prior to Hansen's 1988 paper had ever suggested that tropospheric ozone is an important contributor to the greenhouse effect.

Yet the facts say differently. Between 1960 and 1980, tropospheric ozone increased at a rate between 1 and 2 percent per year, meaning that ozone increased 22 to 48 percent over that 20-year period. During that same period, carbon dioxide increased by only 6 to 7 percent. So the assumption by the GISS general circulation model that ozone will increase by only 25 percent while carbon dioxide and methane concentrations will increase 100 percent is quite unrealistic when it appears that currently tropospheric ozone increases at a rate about three times faster than the rate of carbon dioxide increase.

By using a simpler kind of climate model, one that does not have the detailed three-dimensional structure of a general circulation model, scientists can more easily perform sensitivity studies to isolate the impact that an increase in only one trace gas will have on the greenhouse effect. Most of the models set up this way show that doubling tropospheric ozone affects the earth's climate one-third as much as doubling carbon dioxide. Since the rate of increase of ozone is about three times greater than the increase rate of carbon dioxide, ozone's impact on the earth's climate should be comparable to that of carbon dioxide over the same time period. So one could argue that Hansen's model calculations, especially between 1960 and 1980, underestimate the actual temperature increase because of the lack of correct ozone data.

Figure 24 shows the contribution of each trace gas increase to the computed temperature in the simplified model that is also available at GISS. In this graph the one exception stands out distinctly: the computed increase in temperature due to increase in tropospheric ozone. In the graph the solid bar above O_3 represents how much warmer it would be if tropospheric ozone had increased by 100 percent between 1880 and 1980.

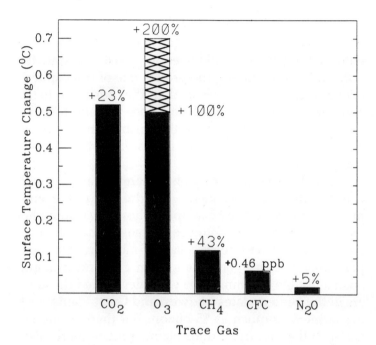

Figure 24. The impact of warming due to increases of each trace gas that contributes to the greenhouse effect. The number above or next to each bar represents the percentage increase of that trace gas during the 100-year period 1880–1980, except for the chlorofluorocarbons (CFC), which were not present in 1880. The rate of increase of tropospheric ozone has been estimated from the Montsouris data and is assumed to have been between 100 percent and 200 percent over that 100 years. All other trace-gas increases have been documented from ice core samples.

The hatched area above it represents how much warmer it would be if the increase had been 200 percent. From data from the Montsouris Observatory outside Paris, an increase of 100 to 200 percent during the past century is reasonable, but how representative these data are of the entire hemisphere is wide open for debate. Despite the difficulty of quantifying the precise rate of tropospheric ozone increase, it is clear that the northern hemisphere's increase in ozone levels since the late 1800s has very likely had an impact on the earth's climate system comparable to the impact of increased carbon dioxide concentrations.

But there is more to global change than just high concentrations of trace gases. The drought of the summer of 1988 was a good example of how devastating hot weather can be to agriculture, because what often accompanies hot weather is a lack of moisture. Hot temperatures and dry conditions often go hand in hand. Such a double whammy was what devastated the farmers in the United States Midwest and Great Plains, costing farmers as much as $5 billion. But there is another factor to the picture, a third element much more silent and insidious. While the heat and aridity weaken and stress the crops and vegetation, increased ozone levels move in and damage the plants irreparably. This is what happened in the summer of 1988, a summer not only of record high temperatures and dryness, but also of the highest ozone concentrations ever measured over much of the United States. For example, sites in the Shenandoah Forest in western Virginia had never experienced ozone concentrations above the National Ambient Air Quality Standard until that summer. It's hard to prove conclusively, but it is possible that the yields of many crops were reduced more by high levels of ozone than by the heat and aridity. Nonetheless, the triple threat of drought, heat, and high ozone readings may be a sad portent of the way things will be in the future.

Whether or not the greenhouse effect is already here, as Hansen contended, it is clear that at this point scientists are continuing to argue about it. We take the position that the GISS general circulation model may have underestimated the amount of warming that has already occurred by not properly including the contributions of increases in tropospheric ozone concentrations. Such increases would have affected only the northern hemisphere, because that is the hemisphere where such increases have been observed. Such a disparity between hemispheres would complicate the global picture even more. All the other greenhouse gases shown in Figure 24 would increase about the same in both hemispheres. An uneven greenhouse effect, with the northern hemisphere warming faster than the southern hemisphere, would alter the earth's circulation patterns differently than if both hemispheres were warming uniformly. This kind of scenario has never been seriously considered, even though, if in fact this is what is happening, the implications are important. For example, the average trajectory of hurricanes in the Atlantic Ocean may shift slightly farther north than the current trajectory and would dramatically affect the number of storms hitting the coast of the United States.

USING THE PAST AS ANALOGY

If the earth is truly warming, what will it be like? How can we know just how much warming will take place and how our lives will change? One way scientists speculate on the future is to find a time period in the past with similar characteristics, much as they use raw statistics from the past to feed a computer, which gives them a picture of the future. Meteorologist Harold W. Bernard, Jr., in his 1980 book *The Greenhouse Effect*, examined several past climatic periods, especially a period known as the Little Ice Age, culminating in the 1600s, and an earlier warm era known as the Medieval Warm Period lasting from 1000 to 1200. During both these periods the dominating influence on weather in the northern hemisphere was the prevailing westerlies, or the jet stream. The jet stream is the axis of the strongest westerlies, reaching speeds up to hundreds of miles an hour and confining itself to an altitude of about 30,000 feet. The importance of the jet stream is that it is what pushes weather patterns across continents and oceans. The storms and cyclones spawned in the southwest United States, or along the Gulf Coast, which then sweep eastward and up the Atlantic seaboard, are guided by the jet stream. The jet stream seldom travels in a straight west-to-east line; it undulates and sometimes loops around the globe. Local changes in the weather are dependent on the jet stream's location and configuration. For example, with the jet stream looping down over the middle of the United States, then sweeping up the Atlantic Coast in a northwestward direction, the East Coast could be buffeted by a series of Gulf storms, meaning rainy and moist weather, while behind the loop of the jet stream, in the western half of the nation, dry, cool weather would prevail as the westerlies kept air flowing from the northwest.

Using data from the 1930s, when the jet stream configuration brought hot, dry weather to the middle of the country, we may be able to get an idea of what the United States will be like in the year 2040, when scientists speculate the full effect of global warming will be felt. Although the decade of the 1930s was a period of above-normal temperatures, weather records show that the warming was not universal. While the Midwest and Great Plains states suffered from record high temperatures, generally the eastern third of the nation experienced relatively cool, moist weather. In Boston, for example, the average annual temperature during the years 1930–1939 was actually a degree *cooler* than normal. What this means is that an increase in global temperatures will have effects that will not be universal. Some areas may feel the effects much more than other areas.

But as in the past, as the earth warms, there is a concomitant shift in the westerlies and the jet stream, which will bring about definite changes in weather patterns. How exactly the jet stream is configured will determine what kind of weather will be experienced in certain areas of the country. As in the 1930s, part of the United States might experience drought and heat similar to the dust bowl conditions back then, while other areas may not even notice the summer's heat. But in the winter, those areas may experience warmer weather than usual, while where it was hotter in the summer it may be colder in the winter. What we can say for sure is that the weather will change and keep changing. In fact, the primary effect of a global warming may be the *changeability* of the weather. During the 1930s, one of the warmest decades in modern times, many of the areas that experienced such devastating heat during the summers also experienced intense cold waves during the winter months. During the decade of heat, San Francisco and Los Angeles both experienced snowstorms, and San Francisco experienced its coldest day ever on December 11, 1932. In 1934, the year which proved to be the hottest in the decade, New York State experienced its coldest temperature on record—a bone-chilling −52° F recorded on February 9 at Stillwater Reservoir. Record lows were also recorded that month in Boston (−18°), Providence (−17°), and as far west as Sault Sainte Marie, Michigan, which recorded a low of −37°. So it is likely that a future earth warming will bring about not only more heat and drought, but a more varied mix of weather, weather of a more extreme nature.

Currently such issues remain speculative. Possibly Hansen's statements before Congress were nothing more than speculation. But he was using some of the most sophisticated tools currently available, and as director of a reputable scientific institute, he was certainly knowledgeable. Back in 1978, some scientists who claimed that the increase of tropospheric ozone was due to human activity were labeled quacks. But time and careful analysis proved them right. Time and future analysis will eventually be Hansen's jury, too.

CHAPTER SIX

The Forests of Oz(one)

Ozone's Effects on Forests and Crops

On the morning of September 23, 1988, Dr. Barrett N. Rock, associate professor of forest resources at the University of New Hampshire's Institute for the Study of Earth, Oceans, and Space, waited for a television crew from New Hampshire Public Television. The TV people were interested in his research, and they planned to accompany Rock and his graduate assistant, David Moss, to the Mt. Moosilauke site, where they would tape the scientists "doing their thing."

That didn't bother Rock much. Appreciating the interest in his project, he welcomed a chance to tell the public what he was doing. What did bother him was the time it would take. This meant time away from his research and his teaching work. Though only a day, it was a busy time and part of him rebelled at having to put on a show just for the television cameras. The timing was off, since Moss had already made his final trip to the site three days earlier. Now they would have to load all the equipment into the van, gather the crew, and head for the mountain to do it all again.

Rock is a botanist trained in the study of plant anatomy. He's one of those specialists in the field who's spent his life peering through a microscope at the world most of us never see, the world of cells, protoplasm, and the most minute particles that make up what we call life. The limits of his domain are dictated by technology and by how far the microscope can probe into the cellular structure of plants. Since 1978, his research has focused on the kinds of damage ozone and air pollution wreak on our forests.

As part of that research, graduate student David Moss was conducting a survey of spruce needles collected over the growing season at the Mt. Moosilauke site in the White Mountains of New Hampshire. Throughout the summer Moss had made the two-hour trip to the mountain every two weeks to collect needles from two sites, one at an elevation of 2,600 feet and the other 500 feet higher. The upper elevation site was located next to an Environmental Protection Agency (EPA) atmospheric monitoring station operated by Dartmouth College personnel. The upper site was also characterized by spruce trees showing symptoms of forest decline damage. Rock hoped to quantify the damage, by monitoring the continuing effect of ambient air upon the cells of the spruce needles. Scientists have known that the forests in the northeastern United States and the adjoining forests of Canada have been damaged. Published reports of leaf damage and stunted growth had appeared before, and several research groups were studying the problem at different locations in the Adirondacks and in the Green and White Mountains of Vermont and New Hampshire.

When the TV crew arrived, Rock and Moss loaded the van and led the crew to the site. The day was hazy and warm, still summer despite the equinox just a few days before. David Moss, knowing the route by heart, could probably have driven it wearing a blindfold. He also knew the site so well that he felt he had a relationship with the mountain. This might seem strange to someone unaccustomed to working outdoors on a regular basis, but when one works frequently in the forest, such a feeling is not unusual. Moss loved his work. He knew the trees he came to visit regularly like family. He monitored the same trees every time, came to the same place, pruned the needles, even talked to the trees as he worked. Trees are living things, after all, and quite lovely things at that, he felt.

Forests, especially those in undisturbed areas, are ideal for study because they have been exposed to their environment for long periods of time and have not been dependent upon cultivation. Therefore, changes in the environment may well affect these types of vegetation first because they are not protected from the changes by such tools of cultivation as fertilizers, pH-balanced water, lime, and so forth. Also, forests are exposed to the atmosphere throughout the year, day after day, while cultivated crops are usually annuals, growing for a season and then harvested. Thus forests, according to Rock, may well represent our first "early-warning system" indicating global environmental change.

Damage in eastern U.S. forests is generally worse at higher elevations. There are several reasons: more severe weather, poorer soil conditions at the higher elevations, and higher wind-generated evaporation rates, which limit the amount of moisture available to the tree. The reason for two sites at two different elevations is that Moss can compare the needles at the lower site (the "low-damage site") with needles taken from the higher elevation (the "high-damage site"). Needles from the low-damage site are, in effect, a "control" group, representing the "normal" growth of trees unaffected by the same conditions affecting the trees 500 feet higher. Although 500 feet may not seem like much of a difference, the difference in condition of the spruce trees between 2,600 feet of elevation and 3,100 feet is significant. Rock points out that because of the latitude of these mountains and the way weather moves across them, 3,000 feet is high enough so that the tops of such mountains often confront air systems before they reach the lower elevations.

The TV crew set up its equipment at the low-damage site and began taping. Moss and Rock went about their tasks. They pruned needles from the trees and placed them in vials of a special preserving solution, while Rock explained to the cameras what it was all about. It was a staged event, a replay of a few days before, all done for public television. But that was okay with Rock. Like most dedicated research scientists, those trees and the needles from them were his life, and he enjoyed sharing it with a public that too often didn't understand the importance of this type of detailed scientific research.

Then it was time to trudge up the trail to the high-elevation site. Both sites at Mt. Moosilauke are located off an old logging road, more like a dirt path barely wide enough for a four-wheel-drive jeep to fit through. The trail begins at the base of the mountain behind the Moosilauke Ravine Lodge and meanders upward following the contours of the mountain. A power line cuts straight across the ridge from the lodge to the high-elevation site, and usually, when Moss and his helpers are working on the mountain, they follow the path made by the power line to get from the lower to the higher site. The reason the power line is there is to provide electricity to the EPA/Dartmouth College Mountain Cloud Chemistry Program (MCCP) monitoring station. The purpose of the station is to monitor atmospheric and cloud chemistry at the upper elevations of the mountain.

The MCCP monitoring station on Mt. Moosilauke is staffed by a crew of three or four people around the clock. The small cabin and associated tower are full of sophisticated instruments used to measure the atmospheric and cloud conditions. Because of the television crew's equipment, Moss and Rock led them to the upper site along the meandering trail and not straight up the power line right-of-way. As they trudged up and around the hairpin turn in the trail to the site, one of the MCCP staff members was waiting for them outside the cabin. Rock asked her if anything interesting had happened since their last visit three days before. To Rock's surprise, she appeared excited.

"You won't believe what happened," she said.

"What?" Rock asked.

"We had an event here. A big one," the woman said. *Event* is a general term that could mean any number of things but usually conveyed something of significance.

"Three days ago," she told them, "a cloud came through here. The pH reading was 2.6." (A pH reading is a measurement of acidity, and 2.6 is highly acidic. In fact, the most acid reading ever recorded for a cloud is 2.4. The woman's report meant the mountaintop had been covered with an acid cloud shortly after Moss's last trip.) Hearing that, Rock whistled through his teeth.

The trees under study were located practically in the backyard of the MCCP monitoring station. David Moss didn't need the spectrometer or the microscope to tell him something had happened. He took one look at the trees, trees he had been visiting for months, and knew right away what had happened. While the TV cameras rolled, he pointed out to Rock one of the trees he was very familiar with. It had been the weakest at the site, the one tree that had apparently suffered more over the years than the others. But just three days before Moss had gathered a sampling of first-year needles from the tree. Now, as he pointed out to Rock and the TV audience, the tree was missing most of its first-year needles. Where the needles had been at the end of the branches, now only naked branches and next year's buds remained, nothing else. And that wasn't all. The older second- and third-year needles were now dry and falling off. Making the collection that day was like taking out a Christmas tree that had been in the house too long.

"Wow," Rock said, looking at the damage. "I've never seen anything like this before. Incredible."

Later, explaining the event to a reporter, Rock said, "It's a perfect example of what is happening to our forests. An acid cloud sweeps across from the west, and within a few days the trees have lost most of their needles. What better proof do you need that these forests are in trouble?"

THE DREAM OF THE MACROSCOPE

So perhaps the question isn't "Are our forests in trouble?" but "Just how bad is it?" This is the question Rock asks and has every intention of getting an answer to. The research at Mt. Moosilauke consists of monitoring any cellular changes in the needles of red spruce trees. Red spruce trees were chosen because they are a common species, known to be moderately sensitive to atmospheric changes. But there is something else driving Rock: a dream which may best be described as "the dream of the macroscope." The dream started more than 10 years ago when Rock, then a fledgling professor at Alfred University in New York, accepted an invitation to be the on-site botanist in West Virginia, as part of a study being developed within the joint NASA/Geosat Test Case. At the time Rock, like other beginning professors with a family to support, was only looking for a chance to make some extra money and an opportunity to introduce to his students a part of the country besides New York State. During his involvement with the NASA/Geosat study, which aimed to study the earth with remote sensing systems onboard aircraft and satellites in search of symptoms of buried natural gas or heavy metals, Rock came to realize two things. First, there was a wealth of botanically significant information available through the use of airborne and spaceborne radiometers and spectrometers, devices used to measure reflected light. Second, NASA didn't have scientists trained as botanists, people who could recognize the significant information, in its permanent employ. He applied for a job at the Jet Propulsion Laboratory

(JPL) in Pasadena and was hired by NASA. In 1981, Rock and his family moved to California and he began working at the JPL. He supervised a research project involved with the remote sensing of damage in forests. Rock's job was to monitor the sites on the ground that the aircraft and satellites had flown over and try to characterize a "spectral signature" associated with damaged trees, that is, some spectral characteristic of plants under stress.

The necessary tool for this is the spectrometer. A spectrometer is similar to the vision enjoyed by Superman, who, if you remember, was able not only to see what other earthlings saw, but to see more. Superman could see inside things with his x-ray vision. He could see what was happening distances far greater than the human eye could see. In a way, this is what the spectrometer has allowed scientists to do, extending the limits of human vision.

Human vision depends upon reflected and transmitted visible light. Visible light, however, is only a small part of the electromagnetic spectrum of energy that comes from the sun. This energy travels in waves, and the waves are measured in nanometers (nm), with each nanometer representing one-billionth of a meter. What we can see, the visible portion of the spectrum, ranges from 400 to 700 nm. The way human vision works is to see whatever wavelength is reflected by the object it is seeing, as long as that wave is between 400 and 700 nm. Anything below 400 nm is in the ultraviolet range, and anything above 700 nm is in the infrared range. Neither of these areas is "seen" by humans. If you're looking at a blue book, that book is absorbing all the electromagnetic energy between 400 and 700 nm except that around the 450-nm mark, which it reflects back to our eyes and which we know as the color blue. So what we see is limited to what is reflected and absorbed between the ultraviolet and the infrared limits of the visible spectrum.

The value of a spectrometer is that it can "see" more of the electromagnetic spectrum. A spectrometer can quantify what occurs between reflectance and absorption of electromagnetic waves in the visible and part of the infrared portions of the spectrum. While doing his on-site investigations using field-portable spectrometers, Rock wondered if it wasn't possible to apply the principles of the spectrometer from the air, to use the concept of a *macroscope*—a larger view, just as a microscope provides a smaller view—to study forests. Both instruments extend the capability of the human eye to "see" botanically significant information, so that state-of-health assessments may be made by the use of either instrument. The advantage of the macroscope is that it gives you the "big picture." Such an instrument would enable scientists to study plants from a distance, allowing them to gain a broader view of plant changes than was currently visible through a microscope or with the naked eye. With the technological advances made in recent years, it is possible to obtain high-resolution data from jet airplanes flying at altitudes as high as 15 miles, and even from orbiting satellites. To carry the analogy further, while the botanist at the site can use a microscope with a variety of lenses to obtain differing views of the cellular structure of plants, remote sensing instruments, even from orbit, can refine the data from coarse to fine spectral resolution, much as the lenses of the microscope do. By using the reflectance data from a spectrometer, Rock imagined that if scientists knew what to look for they would have an excellent tool for detecting and monitoring the extent of damage to forests and other types of vegetation.

In the years since he has been studying forests, Rock has become convinced that the damage occurring in the forests can be monitored from space. Although much of the publicity surrounding forest damage has focused on acid rain, Rock and other scientists are becoming convinced that acid rain may be only a part of the problem. High ozone levels are also a key player in our air pollution.

But how do ozone and other pollutants damage plants? And why should we care? Anyone who has had a basic course in biology knows that there are more similarities than differences between plants and animals. After all, as far as we know, in the entire universe, plants and animals represent the two great types of living things. On the surface, the main difference between the two is that plants are able to make their own food from materials taken directly from the external environment. Animals, on the other hand, are unable to perform such a feat and must eat either plant or animal food to survive.

If you look at a tree, then at a gorilla, it is easy to see the difference between a plant and an animal. Both a tree and a gorilla are complex organisms, made up of billions and billions of cells interrelated in a complex system of cooperation. But looking at both the tree and gorilla through a microscope, you would find a remarkable similarity. Both forms of life contain the same basic "stuff of life": protoplasm. Every plant and animal is composed of living protoplasm and numerous nonliving materials which protoplasm has made and distributed in the construction of the individual organism. The difference between plants and animals is in the complexity of the organism and how it makes its food.

Trees have been called the planet's first line of defense—an early-warning system. If we consider the entire planet as a living organism, then to study how the atmosphere affects the life and well-being of trees is to study how the atmosphere affects the other life forms dependent upon it. Everything, whether plant or animal, depends on the same air, the same water. Gazing through his microscope at microthin sections of spruce needles, Rock and his colleagues are studying the basic unit of life: the cell. And to determine how that spruce leaf cell is affected by certain chemicals is to determine how those chemicals—in this case gases such as ozone and sulfur dioxide—affect our own cells.

The effects of ozone on plants have been studied for years. But most of these studies were laboratory experiments. Only recently, when it became apparent both in the United States and in Europe (especially in the Bavarian forests of Germany) that damage was being done, have scientists taken to the field to explore the complex interaction of ozone, sulfur dioxide, and climate and soil conditions with plant life firsthand. But laboratory experiments have given scientists a fairly clear picture of how ozone affects plants. Plants have organs, just as animals do. One of the most important organs in a plant is the leaf. A plant lives through its leaves. It is here where food and nourishment enter and where waste products are released. The leaf is the plant's opening to the vast world, where the subtle and complex interaction called *photosynthesis,* the process of food production, takes place.

The "mouth" of the leaf is the stoma. Stomata are extremely numerous, small, openings located in the inner layer of the epidermal cells, just under the surface. Stomata are so very small that even vast numbers of them still make up a relatively tiny proportion of the total epidermal area. One pumpkin leaf, for example, contains as many as 50 million stomata. Through these stomata pass oxygen, carbon dioxide, and water vapor. Each stoma is protected by guard cells. The guard cells are concave in shape and serve as doors, opening and closing the stoma at appropriate times. The guard cells are quite remarkable in the way they work. They manage to change form to open or close the stoma. This change occurs when the cells gain or lose water, a process of swelling and distension. When water pressure is increased, the cells become distended and the shape becomes more curved, thus enlarging the opening of the stoma. When water pressure is decreased, the shape of the cell becomes straighter and the opening of the stoma is closed. The turgidity, or pressure, of the guard cells is dependent, among other things, upon photosynthesis occurring in the guard cells. When photosynthesis takes place, the guard cells become turgid and the stomata open. When photosynthesis ceases, the cells lose water and the stomata close. Thus, during daylight hours the stomata are generally open, while during the night, when photosynthesis isn't taking place, they close. As long as the stomata are open, water vapor is lost from the leaves, and in the process, water is taken in through the roots to replace the water lost through evaporation. Thus stomata control not only gas exchange, but also water movement in the plant.

To understand how ozone affects plant cells, it's important to understand the dynamic process a plant uses to feed itself and grow. Plants absorb carbon dioxide from the atmosphere through the stomata and transport water from the soil through the roots and stem to the leaves. Beneath the epidermal cells lies another layer of tissue composed of cells called the *mesophyll*. These cells contain chloroplasts, which react with sunlight to convert the carbon dioxide and water into carbohydrates. This is the basic process of photosynthesis, and it occurs only in the mesophyll of the leaves, except for the minor amount which occurs in the guard cells. At the same time, the plant absorbs the water and mineral nutrients it needs for growth from the soil. So for a plant to be healthy, it must bring together elements from the air and from the soil. When either one of these sources is polluted, the health of the plant suffers.

In laboratory studies, when a plant is exposed to elevated levels of ozone, the gas affects the chloroplasts in the mesophyll cells and thus decreases the process of photosynthesis. When exposed to ozone, the guard cells become less turgid and the stoma closes as a protective measure, further decreasing the photosynthesis process. What this means to humans is that trees suffering ozone damage aren't as healthy and green as they should be and often, if the damage is severe enough, lose their leaves before the growing season ends. In Europe they call this phenomenon the *early autumn syndrome*. But the damage isn't always visible. In fact, recent experiments using low levels of ozone show damage that is not as obvious, but just as damaging in the long run, as damage that occurs with high doses of ozone. Although the dynamics still aren't completely known, it appears that the ozone in ambient air does its damage by diffusing into the sensitive cellular protoplasm of the mesophyll cells and affects the photosynthetic capability of the chloroplasts. Thus, much of the damage to plants from ozone occurs before there are visible signs.

When ambient air with high levels of ozone first enters the stomata of a leaf, the ozone in that air is a gas which quickly dissolves in the aqueous layer of the cells lining the air spaces below the stomata. Once dissolved in the liquid water, the ozone diffuses through the cell wall into the mesophyll cell membranes, where it affects the individual cells. That's why the damage to leaves from ozone and other pollutants, at least in the early stages, may not be visible to the naked eye. And not all trees in a given area will suffer equally from the ozone in the air. In healthy trees, it is believed, the guard cells close the stoma when exposed to ozone. But by closing the stoma to defend itself, the tree is interfering in the photosynthesis necessary for health. Scientists speculate that this process may account for the wide variability in how plants react to ozone levels, why some species and even some individuals within those species suffer more damage than others. The tree's defense system could protect it from short doses of high levels of ozone, but over the long run even healthy trees and plants suffer if elevated levels of ozone occur over periods of hours to days. Damage from ozone exposure is also dependent upon other factors affecting the general health of the plant, such as soil conditions and available water. This combination of factors may be one reason why most of the visible forest damage occurs in the higher elevations of mountain ranges, where the soil and moisture are often marginal.

Sorting out where the most damage occurs is where high technology can help. With the aid of advanced instruments that allow scientists to study the spectral signature of the forests from aircraft or spacecraft, Rock and his colleagues hope to be able to determine just which forests are in danger. To do this requires cooperation between the aerospace industry and the scientists in the field. In the field in New Hampshire, David Moss and his helpers go about their tasks with enthusiasm. They spend the growing season collecting samples, and they spend the winter studying them. Every two weeks during the summer growing season, Moss loads up the university's van with his gear: a specialized spectrometer, called a *visible infrared intelligent spectrometer* (VIRIS); glass vials containing fixing solution for the needles; pruning shears and a long-pole pruner; and other assorted tools needed in the forest. At the low-damage site on the side of Mt. Moosilauke he and his crew collect samples, examining the same trees every visit. First, they cut a branch containing needles from the first, second, and third years of the tree's growth. Then they carefully select a segment from the middle of each needle and quickly put it into a fixing solution of formalin–acetic-acid–alcohol (FAA), so the needle is "frozen" in its lifelike state, as when it was cut. Then they walk up the path, past the MCCP station, and collect similar samples from the trees at the high-level site. At each site they also collect needle samples for spectral measurement with the VIRIS.

The process takes a full day, but that is only the

beginning. Back at the laboratory, Moss places the needle segments in a vacuum for 30 minutes and then in a solution of 70 percent ethanol to store them. The needle segments are then run through a series of alcohols, a procedure continued sometimes for weeks, until the needles are totally dehydrated. Then they are placed in a paraffin matrix, embedded in a medium that will allow them to be sliced tissue-paper-thin and studied under a microscope. This is the basic process that then allows Rock and Moss to see the individual cells inside the needles and determine the damage done to them on the slopes of Mt. Moosilauke. The damage seen at the cellular level is similar to that seen by the naked eye in trees that have been exposed to ozone and other air pollutants.

That takes care of the microscopic aspect of Rock's research. For the other, the macroscopic view, a company from Canada, the Moniteq Company, Ltd., supplies the aircraft and an airborne spectrometer that make flyovers of the Mt. Moosilauke site to coincide with the ground samples Rock and Moss collect. The airborne spectrometer is an instrument known as the *fluorescent line imager* (FLI), an imaging device that provides a spectral signature for every point on the ground that it "sees." Thus, by comparing the spectrometer readings taken on the ground by the VIRIS with those taken in the air by the FLI, and comparing them with what appears in the microscope, Rock hopes to identify a spectral signature, a signpost common to both the spectrometer and the microscope and diagnostic of forest damage. Already he has discovered such a signature in the needles, but the problem has been to determine whether or not such spectral signatures can be detected at the canopy level of vegetation by airborne spectrometers such as the FLI, as well as by those he envisions beaming down to earth from satellites. The studies in New Hampshire are too new to allow any definite conclusions, but Rock's spectral signature project results, from the time when he was with NASA's Jet Propulsion Laboratory, have already been published, and he expects the results of his current study to confirm the ability of airborne spectrometers to detect the signatures characteristic and diagnostic of forest decline damage.

In an earlier study, Rock, with fellow scientists T. Hoshizaki, J. Vogelmann, A. Vogelmann, S. K. S. Wong, and others, first gathered needles from two sites, one at Camels Hump in the Green Mountains of Vermont, the other from the Black Forest region of Baden-Württemberg in West Germany. They gathered needles from spruce trees at two different elevations, typifying high-damage and low-damage conditions characteristic of each area. These needles were then spectrally characterized with the VIRIS and also studied under the microscope for pigment and cell damage. The FLI was flown over both sites. The FLI readings and readings from the thematic-mapper device onboard the Landsat satellite were compared with field VIRIS readings from both sites. After an analysis of the data, which included poring over highly technical readouts from the various spectrometers, Rock and his associates detected similar spectral signals from both sites. Since spectrometers work on the basis of reflectance (i.e., how much energy is being reflected back from the object studied), and since the amount of chlorophyll in a leaf or needle may be used as a sign of a healthy or damaged tree or plant, the scientists, by comparing how much energy and at what wave-length the spectrometer was recording, could calculate the apparent health of a tree. The fact that the spectral signals were similar suggest that damage conditions in the cells of the needles were similar, a fact confirmed with the microscope. The dream of the macroscope was on the way to becoming reality.

With such information, Rock foresees a future where NASA's Earth Observing System (EOS), planned for the late 1990s, will carry high-resolution spectrometers such as the moderate-resolution imaging spectrometer (MODIS) or the high-resolution imaging spectrometer (HIRIS) to feed data routinely to ground-based scientists such as himself, allowing them to monitor the health of the world's forests.

EFFECT ON CROPS AND FOOD

But forests and wildlands are only a part of the world's vegetation. Humans have been cultivating crops for thousands of years now. You might say that civilization as we know it is dependent upon the health of our crops. History is full of wars begun over the presence or lack of food in the form of some basic staple such as rice or wheat. And since the 1950s, we have become more aware of how dependent the separate links of the food chain are on one another. When any of those units of the food chain are disturbed, the ripple moves up the chain until it gets to the top of the list: human beings. We all know how a drought may push the price of bread and meat higher, how an oil spill in fishing waters affects the price of fish at our local market, and how a war in Iran chokes off the supply of almonds.

In recent years many scientists have expressed concern that because of the poor ambient air quality, our crop yield is being reduced. Estimates of this loss in dollar terms range from $1 billion to $5 billion per year. Back in 1980, the EPA became concerned about the trend they were seeing, the trend of higher concentrations of ozone spreading out to the countryside and affecting even rural areas. So they set up the National Crop Loss Assessment Network (NCLAN) to estimate the magnitude of national crop loss caused by air pollution. The primary objectives of NCLAN, as spelled out in the EPA reports, are to

> define the relationship between yields of major agricultural crops and ozone exposure and to provide data for economic assessment and the development of National Ambient Air Quality Standards, to assess national economic consequences of agriculture's exposure to ozone, and to advance the understanding of cause and effect between crops and pollutant exposure.

Basically NCLAN is a network of field sites, usually located near already-existing research centers such as universities. The crops are grown on these sites under conditions approximating standard farming practices. But in this case there is a difference. At the field sites the plants are exposed to pollutants such as ozone via the use of open-top chambers. These chambers are essentially upright cylinders with a clear plastic film around the sides. Such chambers provide for doses of ozone that can be manipulated to conform to ambient air concentrations. Within these chambers plants are exposed to a variety of ozone concentrations. Researchers determine how much ozone each group will be exposed to based on the daily variation in the ambient ozone concentrations at each site. In other words, they use the ambient level of ozone as a baseline to determine the ozone dosages for the experiment. The lowest ozone concentration used (by means of charcoal-filtered air) is the control dosage, and it runs from 20 to 50 percent of that in the ambient air. Other dosages are coupled to ambient air levels at the site, so that the days with the highest ambient-air concentrations of ozone are also the days on which the plants are exposed to the highest controlled dosages of ozone. The results of crop growth at the research sites are compared to the results of control groups of plants in nearby fields exposed only to the ambient air and not to controlled dosages of ozone.

The EPA has released a five-volume report on some of its findings, including the results of many tests taken by NCLAN. This particular research is so new that for some questions there are just not enough data yet to allow answers. For example, the answers to whether ozone exposure increases a plant's susceptibility to disease are inconclusive. Certain trees, notably the eastern white pine and the Ponderosa pine, have shown increased disease susceptibility after exposure to ozone, as have potato plants and barley. Yet cabbages and tomatoes have seemed to show less disease after exposure.

Perhaps the most significant results of the EPA and NCLAN studies over the years relate to the acid cloud that Rock and his colleagues saw in New Hampshire. Most episodes of air pollution are not confined to one pollutant. Ozone and sulfur dioxide are the two most common air pollutants—ozone because it is created photochemically, and sulfur dioxide because it is one of the main waste products of the burning of fossil fuels. For years scientists have been experimenting to discover how the interaction between sulfur dioxide and ozone acts upon plants, to determine if the two gases together induce greater damage than one gas acting alone. Two categories of plant responses are possible to the two pollutants. When one pollutant has no effect on a plant but the other one does, then the response is labeled *no joint action*. And the term *joint action* signifies that both pollutants have some effect on the plant. Joint action can be further divided: an additive response occurs when the effect on the plant of pollutants A and B together equals the effect of either one separately. An interactive response occurs when the effect of pollutants A and B together is not equal to the effect of either pollutant separately. We can further divide the interactive response into two categories: synergism occurs when the effect of pollutants A and B is greater than either one alone, and antagonism occurs when the effect is less than either one alone.

One of the first studies to examine the effect of ozone and sulfur dioxide together was done in 1966 by H. A. Menser and H. E. Heggestad, who used sensitive "Bel W3" tobacco plants. In this experiment the plant was exposed to mixtures of ozone (.03 ppm) and sulfur dioxide (.25 ppm) for two or four hours. The plants sustained 23 percent foliar (i.e., leaf) injury after two hours and 48 percent after four hours. Since that study, tobacco plants have been used often in studies of the effect of ozone and sulfur dioxide because of the plants' apparent sensitivity to those two pollutants. In the 1966 study, when the plants were exposed to either pollutant alone for the same amount of time, no apparent visible injury was produced. A later study by D. T. Tingey and others exposed species of alfalfa, broccoli, cabbage, radish, tomato, and tobacco to varying concentrations of ozone and sulfur dioxide, and in all cases the response of the plants was either additive or synergistic. Synergistic responses have also been reported in studies using apple, grape, radish, cucumber, and soybean plants. Further results of the EPA studies show that:

1. When concentrations of ozone and sulfur dioxide are below or at the threshold for visible injury (i.e., changes in leaf color or texture that can be observed with the naked eye), synergistic interaction may occur.
2. As concentrations of ozone and sulfur dioxide increase in mixture above the injury threshold, the yield loss (measured by the actual weight and density of the plant) by joint action may be additive.
3. When both pollutants are present in high concentrations, the joint action of ozone and sulfur dioxide may be antagonistic, so that further weight loss is minimal.
4. In field studies, the addition of sulfur dioxide generally doesn't influence ozone response unless the concentrations and exposure frequencies of ozone are much greater than those of sulfur dioxide typically found in the ambient air in the United States. Conversely, as levels of both gases rise, the effect on vegetation can be serious, as evidenced in the "event" on Mt. Moosilauke.

How these results translate into economic gain and loss is difficult to assess. Obviously, a farmer whose crop of soybeans produces less than it did, say, 20 years ago could justify anger at the climate producing such changes. But such losses are hard to determine because of all the other factors that go into the development of agricultural crops: soil conditions, weather, moisture, plant hardiness, and sunshine levels. The typical farmer is probably unaware of the insidious nature of ozone pollution. But others are suffering also. According to the EPA studies, many people make their living growing ornamental plants such as Christmas trees, wreaths, and indoor plants. One of the first signs of ozone damage in many of those plants is a discoloring of the leaves, and sometimes even a complete loss of the foliage on the plant. Obviously, if one is dependent upon the foliage of certain plants to make a living and that foliage is suddenly, or not so suddenly, tampered with, then economic loss occurs.

The increasing amounts of ozone in the ambient air are costing us money. Most people would probably agree that money is important, but not the most important thing in life. Health ranks right up there, and current studies are showing that the effects of the increasing ozone levels on human health should also worry us.

Smoke Gets in Your Eyes

Effects of Ozone on Human Health

CHAPTER SEVEN

Smoke Gets in Your Eyes

Effects of Ozone on
Human Health

A SHORT HISTORY OF BAD AIR

On the morning of October 27, 1948, residents of Donora, Pennsylvania, awoke to find a thick fog enveloping the river valley. One observer noted that while watching a passing train belch smoke as it started up the grade that led out of town, he saw that the smoke coming out of the locomotive's smokestack was not rising as it normally did. Instead, it oozed over the lip of the smokestack and sank down like a puddle of black ink falling into a pool of water.

Part of the problem was the unusual weather. During the last week of October 1948, the northeastern United States was in the midst of a stagnant weather pattern. A stationary high-pressure cell placed a dome of still air with little or no wind above the area. This meant that as air at the surface warmed and tried to rise (what is called *convection*), it ran into the stagnant, warmer air above it. The result was a temperature inversion, which forced the surface air to remain at the surface and generally prevented any kind of movement.

Days of fog and haze characterized the daily weather picture from Virginia to New England. In the industrialized areas, this combination of temperature inversion and stagnant air created smog, composed mostly of sulfur dioxide and smoke from the burning of soft coal. Especially hard hit were the industrialized river valleys of western Pennsylvania, where because of the mountains the pockets of air in the valleys became saturated with pollutants.

Donora, Pennsylvania, is located on the inside of a sharp horseshoe bend of the meandering Monongahela River, about 30 miles south of Pittsburgh. The plain abutting the river at that point is narrow, and back in 1948 the area along the Donora side of the river was occupied by a large steel mill and a zinc reduction plant. The plain rises sharply on each side of the river, reaching on the Donora side a height of 400 feet in a 10 percent grade.

Part of the problem that day when the observer noticed the train's smoke going nowhere was the inversion layer and the stagnant air trapped in the valley. But another part of the problem was the pollutants being fed into such a weather pattern by the steel mill and zinc plant. The smoke released by these factories was enough to obscure visibility and force residents to turn on their lights in the middle of the day. The condition persisted until October 31, when a light rain began falling and the weather system causing the inversion moved to the east.

What is remarkable about what happened in Donora in 1948 is that during those four days of stagnant weather, a total of 17 deaths occurred. That was 15 more than the average 2 deaths normally expected for the same amount of time. Hospitals and doctors were inundated with calls and visits from residents experiencing coughs, watery eyes, pains in the chest, and other symptoms of respiratory illness. The U.S. Public Health Service (USPHS) was called in, and by December the USPHS and the state health service had undertaken a study to find out the causes of what became known as the Donora disaster, one of the worst cases of air pollution in the history of the United States. The toxic smog affected almost 6,000 residents, which represented over 40 percent of the population of the city. Most severely affected were the elderly; 60 percent of the residents over the age of 60 were reported ill. Of the 17 deaths caused by the incident, the majority were also among elderly, ranging in age from 57 to 84, with a mean age of 65.

Four years later an even greater disaster occurred, this time across the ocean in the city of London. There, from December 5 to December 9, 1952, a high-pressure cell stalled over the British Isles and created the same kind of fog and temperature inversion experienced in Pennsylvania. This was not the first case of such a weather pattern in London, but it became one of the best documented and was probably the most severe case of air pollution in this century.

During the 1952 episode, within 12 hours of the onset of air pollution an inordinately large number of people in the Greater London area began experiencing symptoms of respiratory tract illness. Most of the casualties appeared to be people with previous histories of respiratory illness, with men outnumbering women. Although there is no way of knowing for sure, officials estimated that about 4,000 deaths could be attributed to that particular air-pollution incident. Though the deaths were not limited to any one age group, statistics gathered after the incident pointed to two groups who suffered the most casualties: those over 45 years of age, and those under the age of 1 year—the old and the very young. This incident was disastrous enough to spur the population of London to press for action. The result was the Great Britain Committee on Air Pollution, which the British government formed and empowered to take steps to prevent any similar occurrences.

Though these historical incidents in Donora and London have taken their place in the Bad Air Hall of Fame, they were not the first, or the last, incidents of air pollution to cause widespread suffering and death. Smoke has been both a boon and bane to humanity since the cave people discovered fire. Without fire it is doubtful that the human race would have survived. However, the waste product of fire is smoke, and smoke, when it has no place to go, can be deadly not only to human health but to plant and animal life as well.

Air pollution has been around at least since Roman times. Seneca reviled the "heavy air of Rome," and Roman politicians complained of soot dirtying their togas. During the Middle Ages, Europe began using coal for fire instead of wood, as wood was becoming scarce. Coal creates more pollutants than wood, especially sulfur dioxide, which until the 1940s was the primary antagonist in episodes of air pollution. As far back as 1306, King Edward I of England decreed that no coal be burned during sessions of parliament, because of the smoke it produced. England and the rest of the British Isles, because of the tendency of the weather to produce long stretches of stagnant inversion layers, have long been accustomed to bad air. In the 17th century, the diarist John Evelyn wrote that a "hellish and dismal cloud" enveloped the city of London, and that "catarrhs, consumption and coughs rage more in this one city than on the whole earth besides." London, after all, is where the term *killer smog* originated, referring to the aforementioned disaster in 1952, as well as to a later one in 1962 which claimed 400 lives.

But London was not alone. Here on this side of the Atlantic, the cities of Pittsburgh, St. Louis, and Cincinnati all experienced episodes of air pollution serious enough to spur their governments into action. Before 1940, the nickname for Pittsburgh was "Smoke City," and as late as 1945 Pittsburgh led the nation in number of deaths from pneumonia. But these episodes are past history. Local, state, and national governments have been provoked by the citizenry into taking action. In 1963 the Clean Air Act was passed, giving the federal government limited enforcement power over interstate pollution. Scientists quickly began studying the problem and came up with solutions: burning cleaner fuels and fitting smokestacks with "scrubbers" to filter out particulates. And these solutions worked. By the 1950s, the cities of St. Louis, Pittsburgh, and Cincinnati had changed from burning soft coal to burning hard coal or gas, and had cleaned up their industrial smokestacks, and the pall that had hung over them for years was finally lifted.

But not all air pollution is the same. The kind of air pollution that devastated Donora, London, and Pittsburgh is known as a *reducing pollution,* caused by the reduction of smoke to its particulates. The smog that still plagues Los Angeles and our larger urban centers is known as an *oxidizing* type, which is pollution caused by the oxidization of gases in the atmosphere. What that means is that times have changed, and though the air may not be bad in the same way it was back in the 17th century in London, or in 1948 in Pennsylvania, the overall problem is still with us. Bad air is still affecting our health, whether it is the "killer smog" of London's smoke-filled atmosphere or the brown haze that covers the Los Angeles basin.

THE MACROSCOPIC TO MICROSCOPIC VIEW

Bad air. What does it mean? One way to gain an understanding of or a new perspective on air pollution is to see it from a macroscopic, or universal, view. As far as we know, the universe is so extensive and so vast, that the human mind has trouble grasping its size. As one example of the size of the universe, the *Voyager* spacecraft that left earth in 1977 is just now reaching the further edges of our solar system. Sending photographs back from Neptune in 1989, at the speed of light, took *four hours*. So in the vast scheme of things, the earth may be just the tiniest dot, and yet it is the only tiny dot out of all those millions of dots that supports life, as far as we know. The primary requirement for life on this planet is the particular mixture of gases that hugs the earth's surface—what we call air. The gas mixture is relatively simple: about one-fifth oxygen, almost four-fifths nitrogen, and the rest argon, trace gases, and water vapor. A simple mix, but crucial. For example, if we are deprived of oxygen for as little as six minutes, we will die. The right amount of oxygen in the air for human survival occurs only in the lowest several thousand feet, and most of the air we breathe lies in the shallowest seven feet of air closest to the ground. Anyone who has climbed a mountain knows what happens when the body is deprived of the "right" amount of oxygen—what we call altitude sickness, characterized by headache, nausea, and general malaise.

The agent between us and the air—the medium, if you will—is the lungs. It is through the lungs that we take in the air and give off the carbon dioxide left over from our internal cleansing process. The lungs are quite an amazing feature of our bodies. They are at work constantly, and along with the heart they make up our primary life-support system. The lungs are able to cleanse themselves of most impurities found in the air by the use of marvelous cleaning devices called *cilia.* Without such devices, the lungs would quickly fill up with lethal dust and would stop working. Cilia are small, hairlike projections, actually pieces of tissue, growing on the surface of the windpipe and the bronchial tubes. The cilia beat upward, undulating regularly like a field of ripe wheat in a stiff breeze, pushing along a continuous stream of fluid, called *mucus.* This mucus carries impurities up and out of the respiratory tract to the throat, where it is swallowed and eliminated by the digestive system. A very efficient design, where everything has its place and most contingencies can be easily handled.

The respiratory tract resembles a tree placed upside down in the human chest. The trunk is the windpipe; the branches are the lung's bronchial tubes, called the *bronchi.* At the end of the bronchi are the equivalent of the leaves of a tree, an abundant array of little sacs, called *alveoli*. It is here where the oxygen and carbon dioxide are exchanged, where gas meets liquid, air meets the blood. These tiny air sacs are stuffed and convoluted to fit into the chest cavity, but if an average person's lungs were stretched out and laid flat, they would cover the length of an entire football field. It is in these deep passages, the dark labyrinth of the diffusion centers of the lung, where the real work takes place. The entire blood supply of the body flows through this area, an area that probably resembles the most remote stretches of the Okefenokee Swamp. Here you can imagine poling a skiff along waterways that diminish in size, becoming so narrow that soon the shore and river merge into wet bogland. So it is with this deepest level of the lungs, where the airways diminish in size from the superhighway of the windpipe to the narrow land of the alveoli, growing ever smaller and smaller until alveoli and the blood-carrying capillary merge.

If impurities have evaded the cilia and traveled as far as the alveoli and the diffusion area of the lung, they are not so easily cleansed from the air sacs. Bacteria invading the alveoli are routinely ingested by special cells called macrophages and dissolved there by enzymes. But particulates are a more difficult problem. Some particulate matter can find its way into the air sacs and then into the lymph channels (which make up the diffusion area of the lungs) and remain there indefinitely. The walls of the air sacs are as thin as paper. When they are injured and break down, because of oxidizing agents, particulates, or old age, breathing becomes more and more of a chore. Eventually, the owner of those lungs finds it impossible to get enough oxygen from breathing, and he or she suffers from emphysema.

Sometimes the walls of the bronchial tubes become inflamed, causing coughing and an increased flow of mucus which we call *bronchitis*. Either way, the clean, efficient work of the lungs suffers.

The medical community has known for a long time that bad air causes problems in humans and animals. The disaster in Donora in 1948 and the killer smog of London in 1952 were both well documented and served to kindle an interest in the effects of air pollution on human health. However, at the time of those tragedies, the main culprits of air pollution were particulates (solid particles suspended in the air that managed to enter the lungs during breathing) and sulfur dioxide, a gas produced by the burning of coal. In the case of London, not only in 1952 but in other smog attacks that recurred well into the 1960s, the primary cause of the bad air was all the home fires. At the time, Londoners, unlike their counterparts in U.S. cities, heated their homes with coal. And they burned the coal in open grates. This combination of coal and open fires produced so much smoke in a city of 8 million, that when weather conditions existed where the smoke had nowhere to go (i.e., up), thousands of people died. Even without the bad air spells that gave London its reputation as the "cradle of air pollution," the ambient air on any given day was bad enough so that even as late as the 1970s, when the British had finally developed central heating and had begun to burn coal cleanly, London still had one of the highest rates of bronchitis. In fact, bronchitis was commonly known as the national disease.

Most of the research kindled by the air pollution tragedies in the middle of this century led researchers to explore the relationship between the lungs and the kinds of pollution so common at that time: particulates and sulfur dioxide. The lungs boast a built-in system for filtering out particles that may cause damage to lung tissue. This system involves the upper respiratory tract, i.e., the nasopharyngeal area of the nose and throat. When air is breathed in with particulates in it, the air rushes through the nasal passages at a high velocity. The nasal passage bends sharply where it meets the throat in almost a 90-degree turn. What this sharp bend in the air passage does is take advantage of a basic law of physics: Matter moving in one direction will tend to continue in that direction. This is known as *momentum*, or *inertia*. So when particulates are inhaled, they tend, rather than make that sharp turn downward toward the throat, to hurtle straight ahead where they slam into the mucus-lined walls of the nasal passages. The mucus is then swallowed and the particulates—dust, tiny particles of coal, or other minerals—are given to the digestive system for disposal.

It's been estimated that the nasopharyngeal area collects up to 80 percent of the particulates inhaled. The other 20 percent that manage to get through are, as mentioned earlier, taken care of by the cilia, which pass them up the windpipe much as boisterous students pass bodies up the grandstand during college football games. Any particulates that make it even further are expelled by the coughing mechanism. If the lungs fail to rid themselves of particles and objects, and the particulates make it to the deeper recesses of the bronchi and alveoli, then inflammation may develop and we suffer upper respiratory disease, like bronchitis, emphysema, or pneumonia. Air pollution doesn't cause just one particular disease, which is why it has been difficult to

determine exactly the relationship between lung disease
and bad air. Although correlations do seem to exist between areas of known air pollution and rates of bronchitis and even lung cancer, medical scientists have
been unable to discover a set of symptoms attributable
to air pollution alone.

NEW POLLUTION

Up until the 1960s, when physicians talked about
air pollution they were talking about the obvious kind:
the thick industrial wastes pouring from smokestacks in
the U.S., or the dark clouds above the millions of private residences in cities like London where they burned
coal. But after World War II a new type of air pollution
began appearing. This pollution was different from the
type previously known and was the result of progress
and history. The cradle of this type of air pollution,
labeled by scientists as *oxidizing*, was the city of Los
Angeles. Los Angeles is not London. In fact, it would be
hard to imagine two cities more different. London has
been around for thousands of years and is part of the
Old World. Los Angeles has existed barely a century.
London is a city with a notorious climate: days upon
days of fog and drizzle with cool, sea-swept temperatures. Los Angeles enjoys one of the world's most
comfortable climates: warm days, cool nights, and
abundant sunshine. London is a city that is host to
thousands of factories and mills. Los Angeles has little
manufacturing industry, having built its economic base
on the service industry and aerospace. London has a
quick, efficient mass-transit system that links the city

with its suburbs. Los Angeles has a system of super-highways that carry automobiles and their occupants over and under its sprawling environs.

It is these freeways, as we noted earlier, these long, snakelike ribbons of concrete, that have caused the problem of Los Angeles' notorious smog. Unlike London, Los Angeles' air problems are not caused by particulates and gases released by home heating fuels and industrial waste. The primary culprits of pollution in Los Angeles are the automobile and the abundant sunshine the city has become famous for. The action of the sunlight on the emissions from the millions of automobiles, combined with a geography that produces days of temperature inversion layers, produces a unique "oxidizing" type of smog, the main ingredient being ozone. Scientists didn't begin to study ozone's effects on the lungs until the 1960s, because they had not yet identified ozone as the primary irritant in Los Angeles smog.

Photochemical smog is no longer only Los Angeles' problem. After the air pollution tragedies of the mid-century, which precipitated the Clean Air Act of 1963, cities such as Chicago, New York, Pittsburgh, and St. Louis made changes which did reduce the emissions of particulates and sulfur dioxide from industrial sources and home heating. But at the same time, these and other cities experienced the same enormous growth in automobile use as Los Angeles. Most of the major cities in the United States have seen their downtown areas transformed by expressways and freeways, to enable the thousands of commuters to travel to and from the suburbs. The automobiles used by such commuters spew the same gases into the air as they do in Los

Angeles, with the result that photochemical smog in such cities has become so commonplace during the summer months that the public assumes "urban air" to be of dubious quality. In fact, the air over most of our urban areas during the summer months is often above the National Ambient Air Quality Standards (NAAQS) for ozone, and even on days when it doesn't exceed the level set by the EPA it is still too high to be considered healthy. Thus "urban air" took its place among the other learn-to-live-with-it qualities of big city life such as crime, noise, and crowded restaurants.

At first scientists began to study ozone's effects on human health to determine the danger level of ozone in the ambient air. At what point did ozone do harm to the most people? The research undertaken to find the answer to these questions involved subjecting animals to high dosages of ozone over relatively short periods of time, and applying the results to humans. Thus scientists finally settled on a concentration of 0.12 parts per million as the standard. Above that level, in subjects at rest or under moderate exercise, ozone exposure would bring about symptoms of reduced lung volume, watery eyes, and bronchial constriction. The tests to determine the standard, however, involved exposing subjects to such levels of ozone for only a short period of time. This length of exposure was thought to be adequate for it was based upon mechanistic studies on the diurnal (daily) formation and decay of ozone, which suggested that the highest concentrations would occur for a short time period beginning around midday or in early afternoon. Indeed, this is the pattern in such places as Los Angeles, where the proximity to the ocean usually ensures a cooling, cleansing sea breeze during the mid to late afternoon hours on an average day. This influx of cool air is often enough to move the ozone and break down the inversion layer that causes the severe episodes of pollution.

But in the early 1970s further research revealed a different dynamic for ozone production. In addition to midday peaks, ozone will often accumulate to high levels at various times depending upon certain factors: weather conditions, geography, and the quantity of organic hydrocarbons and nitrogen oxides which make up the primary precursors for the production of ozone. What this meant was that the Los Angeles model of ozone production was not the only one. In 1980 George T. Wolff and Paul J. Lioy, under the auspices of the interstate Sanitation Commission, published a study of the distribution of ozone during three episodes that occurred in July 1978. On July 12 a high-pressure system formed over the Gulf of Mexico, affecting the entire southeastern United States, extending from western Texas northeastward through Illinois and all the way to the Atlantic seaboard. This system created a strong southwesterly airflow pattern taking air over the Louisiana–Texas area and depositing it over the mid-Atlantic coast. According to the researchers, this pattern produced an ozone "river" stretching from Texas to southern New England. This ozone river contained ozone in amounts up to 0.12 to 0.13 ppm, and it persisted for *one week*.

On July 21 a cold front marking the leading edge of a high-pressure system came down from Canada and moved off the Atlantic Coast by the morning of the 22nd. This air mass replaced the warm air of the previous week over all the eastern half of the country except the Southeast. In the South, the air mass of the previous week persisted, the heat wave continued, and ozone levels as high as 0.13 were reported over eastern Texas, Louisiana, Oklahoma, and Arkansas. On July 23 the new high-pressure system moved into Pennsylvania, and the cold front dissipated from Oklahoma to North Carolina. Because movement of energy is always from the hotter to the cooler (remember the reason a hot cup of coffee cools rapidly in the cold air), without a sharp demarcation between the cold and warm air, the air over the South with its high ozone levels began to intrude into the high-pressure circulation pattern. The result was that the ozone levels that had been confined to the South soon turned up in southwestern Michigan and Ohio, ending up in the Northeast as the high-pressure system moved off the coast. Thus another "ozone river" was established, carrying ozone from its source in the South, hundreds of miles away, to the large urban sprawl of the Northeast.

In another well-known study of the dynamics of ozone production, researchers[1] showed time-lapsed movement of ozone through the northeast corridor, stretching from New York City to Boston, during one day of an episode in 1974. Using ozone monitors at various sites along the corridor, the researchers concluded that ozone originating in the New York metropolitan area actually traveled to Boston within one day, a distance of over 200 miles. The mobility of ozone pres-

---

[1]W. S. Cleveland, B. Kleiner, J. E. McRae, and J. L. Warner, Bell Laboratories. "Photochemical Air Pollution: Transport from the New York City Area into Connecticut and Massachusetts," *Science*, Jan. 16, 1976, pp. 179–181.

ents a problem in ozone research and in finding a way
to reduce ozone levels to protect health: ozone is a "secondary" pollutant. Unlike primary pollutants such as
sulfur dioxide and carbon monoxide, ozone is not limited in space or time to the area of its source precursors.
For example, when sulfur dioxide was the problem,
back in the 1940s, the solution was as simple as changing to fuels lower in sulfur, which stopped the production of sulfur dioxide at the source. But the ozone problem is not so easy, for ozone is not produced until it
reaches the atmosphere (thus it is a "secondary" pollutant). It is created by a photochemical reaction that occurs when certain gases are exposed to sunlight. As
long as those gases are present and there is sunlight,
ozone will continue to be produced even over a time–
space distance away from the source of the gases.

WHAT IS SAFE?

This fact—that ozone can continue to be produced
even away from the source of the gases—is important
when looking at our attempts to determine exactly what
are safe levels of ozone. In 1970 the Clean Air Act was
amended to provide the federal government with the
authority to set standards for certain air pollutants
which "may endanger public health or welfare." The
National Ambient Air Quality Standards were created
to provide an "adequate margin of safety to ensure protection of public health." Going even further, the legis-
lature mandated that the standards be set at "the maximum permissible ambient air level . . . which will
protect the health of a *sensitive* group of the popula-

tion." Furthermore, margins of safety are to be provided so that the standards will afford "a reasonable degree of protection against hazards which research has not yet identified."

On April 30, 1971, the EPA published in the *Federal Register* both primary and secondary (i.e., based on pollution effects on vegetation, visibility, crops, and other factors not directly affecting human health) standards for photochemical oxidants. These standards were an hourly average of 0.08 parts per million of ozone not to be exceeded more than once a year. Notice the phrasing: an hourly average, not occurring more than once a year. Eight years later these standards were revised downward, to 0.12 ppm. In addition, the EPA made the following changes: (1) they changed the chemical designation of the standards from photochemical oxidants to ozone (O_3); (2) they changed to a daily maximum standard rather than an "all-hours" standard; and (3) they changed the definition of the point at which the primary and secondary standards are attained to "when the expected number of days per calendar year with maximum hourly average concentration above 0.12 ppm is equal to or less than one."

The factor cited as responsible for the revision of the primary and secondary standards for ozone was published research that indicated several things: (1) an adverse health-effect threshold for ozone could not be identified with certainty; (2) the level at which ozone affects pulmonary response and respiratory tissue is as low as 0.15 ppm; (3) asthma attacks seem to increase when ozone levels reach 0.25 ppm and higher; and (4) according to studies, premature aging symptoms were reported in animals exposed to ozone. In effect, the EPA concluded that a primary standard of 0.12 ppm as a one-hour average, not to be exceeded more than one day per year on average, would protect public health with an adequate margin of safety.

Nonetheless, there are several problems with the current standards, according to some scientists studying ozone. The main problem is that setting the standards based on one-hour averages may not be taking into account damage done to human health from longer exposure to lower levels of ozone. Also, setting the standards based on levels affecting only "sensitive" groups, such as asthmatics and elderly people with respiratory problems, leaves unprotected the vast majority of healthy adults and children who may be affected by breathing ambient air with increased ozone levels, even though those levels may not exceed NAAQS standards. Physicians already know that on a day when ozone levels are on the rise, even slightly, more asthmatics and other "sensitive" groups will start coming to them for relief. But what about the healthy people, those who don't already suffer from asthma and other respiratory illnesses? Are these people being affected by the general increase in ozone? Is there a threat to human health posed by increased levels of ozone even though they remain below the federal standards deemed safe? That is the question that has so far been unresolved.

When ozone enters the body through the respiratory system it can be removed by two processes: dissolution and chemical reaction. The rate, amount, and sites of this process of removal are what determine how an individual will respond to ozone and what adverse health effects, if any, he or she will experience. Much of the research studying how ozone is removed from the body has been done on animals, and most of the research is so new that scientists studying the problem are still not sure of the actual dynamics involved. But from animal studies, it appears that ozone and other gases absorbed into the respiratory system are removed in the nasopharyngeal (nose–throat) area. In one animal study, approximately 50 percent of the ozone was removed in the nose–throat area (72 percent for dogs, 50 percent for rabbits). As mentioned above, much of the foreign matter in air is deposited on the walls of the nasopharyngeal area just by virtue of momentum. But gases pose a different problem, since because of their size they can easily make the bends and curves of the respiratory airway. The process by which ozone is removed in the upper tract is still not clear, but scientists believe it is removed by chemical reaction, with the mucus lining carrying it away in much the same way it does particulates.

Where ozone seems to do the most damage is further down the respiratory system, in the area where the bronchiole and alveolus meet. One way to think of the respiratory system is like a vast highway-transportation network. The major road is the airway leading from the mouth and throat; as the air moves through the air*ways*, the passages become smaller and smaller. The smallest of these passages are the bronchioles, usually less than 1 mm in diameter. From here, the movement of air stops, and the air *spaces* begin, where the diffusion and exchange between blood and air occurs. From the bronchiole, air moves to the end of the road, the alveoli via the air ducts, and air sacs. Each alveolus is a hexagonal or spherical air cell separated from the blood-carrying capillary by a membrane only 0.2 to 0.4 mm thick, through which the blood–oxygen exchange takes place.

So it is in the lower tract that ozone appears to do the most damage. Because most of the research thus far has been done on animals, and because the research is so new, scientists still are not clear about the actual mechanism whereby ozone affects human lung tissue. One theory is that ozone oxidizes small molecules in the lung tissue, thus promoting the formation of a type of scar tissue that interferes with breathing. Another theory is that the dominant feature is a change in fatty tissue, with the same result. However, research is still going on to discover the exact dynamics involved.

In a recent study (1988) using human volunteers, scientists set out to determine exactly how ozone affects the lungs of humans. In animal studies, ozone was shown to produce inflammation in the lungs of baboons, rabbits, and dogs. In the 1988 study,[2] 11 healthy, nonsmoking males, ranging in age from 18 to 35 years, were exposed once to 0.4 ppm of ozone and once to filtered air for two hours with intermittent exercise. Eighteen hours later doctors washed out the bronchoalveolar area of the lungs and analyzed the cells and fluid for various signs of inflammation. They discovered two significant signs that the lung's defense system had been engaged: first, they noted an 8.2-fold increase in the percentage of certain cells (polymorphonuclear leu-

[2]H. S. Koren, R. B. Devlin, D. E. Graham, Richard Mann, and William F. McDonnell. *Atmospheric Ozone Research and Its Policy Implications*, in T. Schneider et al. (eds.), Amsterdam: Elsevier Science Publishers B. V., 1989.

kocytes, or PMN) in the total cell population, a sign of lung inflammation; and second, they found a small but significant decrease in the percentage of macrophages after exposure to ozone, revealing that an "invasion" of ozone had engaged the defensive macrophages, the SWAT team of the lung's immune system. The researchers' conclusion was that an acute exposure to 0.4 ppm of ozone for a short interval results in increased levels of inflammatory cells and soluble factors potentially capable of producing damage in the lower airways of human lungs.

Such laboratory results demonstrate that high dosages of ozone will damage lungs, but they don't address the issue of just how much ozone is dangerous. Clean air contains ozone, but in such minute quantities that it does no harm. Where is the threshold? What about the levels of ozone that are commonly found during warm summer days? To answer such questions, a group of scientists[3] conducted an experiment using healthy, nonsmoking volunteers with an average age of 22 years. With treadmill and bicycle ergometers, a baseline was established for each subject's lung capacity during moderate exercise by the following method: The subject takes a deep breath and then exhales with maximum force. The amount of air is then measured, to yield the forced-vitality capacity (FVC). Such a measurement is deemed a good indication of the strength and health of the lungs. Measurements of FVC are given in liters of air per minute (l/min).

[3]D. Horstman, V. McDonnell, L. Folinsbee, S. Adbul-Salaam, and P. Ives, of the Clinical Research Branch of the Health Effects Research Laboratory, EPA. North Carolina.

In this particular study, each subject was exposed
to filtered air with no ozone and was then exposed to air
with increasing increments of ozone: 0.08, 0.10, and
0.12 ppm ozone on separate days. These exposures
were separated by a minimum of one week, and the
exposure sequence was randomized. Neither the subjects nor the experimenters directly involved knew on
which days ozone was present. Upon arrival at the laboratory, each subject was given a brief physical examination. Forced expiratory spirometry (a measurement of
breathing capacity) was used, and the experimenters
evaluated pain upon deep breathing before and after all
exposures. So after the baseline measurements had
been determined, the subjects were put through their
paces, which in this case involved exercising on either a
treadmill or a cycle ergometer at a previously determined intensity. During each exposure, each subject
went through six 50-minute periods of exercise (alter-
nating between treadmill and cycle) and examination.
During each exercise period, the experimenters measured the subject's expired volume per minute (VE),
oxygen consumption (VO_2), and heart rate for 3 minutes
after 40 minutes of exercise.

The results of the study showed conclusively that
even at levels below the 0.12 threshold set by the EPA,
healthy exercising adults show signs of lung damage.
The exercise performed by each subject was planned to
duplicate not the level of the prime athlete but a level
equivalent to a day of the kind of moderate or heavy
work many people perform. In other words, the experiment was designed to measure the effect of ozone on
those subjects performing average work, so the exercise
levels of the subjects were designed to simulate (though

not exactly) the workload of, say, a stock clerk or a warehouse person. The scientific team reported decreases of lung function and increases in airway reactivity (an inflammatory response) in all subjects at all three levels of ozone. Though the responses at the highest level (0.12) were greater than at the others, the lower dosages of ozone still produced marked responses that indicated lung inflammation. In their concluding statement, the researchers stressed the relevance of their study to everyday experiences since the intensity, duration, and metabolic requirements of the exercise were representative of a day of moderate to heavy work or play, and the ozone levels and patterns were similar to the ambient levels and patterns often occurring in many areas of the United States and the rest of the northern hemisphere.

An even more relevant, and perhaps alarming, series of studies[4] was completed over a period of years during the 1980s. It assessed the effect of ambient ozone levels on children at summer camp. Beginning in 1980 at Indiana, Pennsylvania, and in 1982 in Mendham, New Jersey, researchers set up their equipment at the site of day camps in each of these rural areas and set about measuring in the children attending the camps such things as how much air was expelled when their lungs were forced to exhale (forced expiratory volume, or FEV), how much force could be applied with a forced exhalation (FVC), and the highest forced flow during

[4]Dalia M. Spektor, Morton Lippmann, Paul J. Lioy, George D. Thurston, Kenneth Citak, D. J. James, Naomi Bock, Frank E. Speizer, and Carl Hayes, "Effects of Ambient Ozone on Respiratory Function in Active Normal Children," *American Review of Respiratory Disease*, 1988, 137:313–320.

exhalation (peak expiratory flow rate, or PEFR). The results of these preliminary studies led the researchers to do a more complete study, this time at a residential YMCA summer camp nestled in the hills of north-western New Jersey, on the shore of Fairview Lake. The goal of this study, completed in 1984 but not published until 1988, was to definitively establish the effects of ambient exposure to ozone on the respiratory functions of active children. It differed from the two previous studies in its use of a residential camp, where the children were available for respiratory measurements every day. Also, their sleeping shelters had screens but were not fully enclosed, so their exposure to ambient air concentrations was consistent throughout the study period.

As in the other two studies, the location of the camp was far removed from strong local sources of air pollution, though it was within the portion of the north-eastern United States which is periodically exposed to hazy air masses containing ozone and other pollutants, and it was well within the "floodplain" of the ozone river discussed earlier. The study was performed on 91 children in residence at the camp for either two or four weeks. The children included 53 boys and 38 girls, ranging in age from 8 to 15 years. The ethnic composition of the group was 72 whites, 15 blacks, 3 Asians, and 1 Hispanic. The level of exercise included in the camp schedule was another point in the study's favor. It was equal to or even greater than the level of activity performed in the other two studies.

Scientists have two primary alternatives for performing research on the environmental effects of the atmosphere. One is to duplicate the conditions under question in a laboratory, which is called a *controlled study.* In the case of ozone, for example, scientists can inject controlled amounts of the gas into laboratory cubicles and monitor the response of the subjects. For ethical reasons, most of such laboratory studies have involved animals, because ozone is known to be toxic to humans. But controlled laboratory studies have a fault: scientists are never sure if they can extrapolate their findings to the "natural condition" outside the laboratory.

The other method, called *epidemiological,* involves studying a sample of a population that has already been exposed to whatever factor is being studied. During the Donora episode of air pollution, and in London in 1952, scientists were more like detectives, backtracking from the effect to the cause. For an epidemiological study to succeed, there must be enough data from "before" to compare with the data "after" in order to find conclusive evidence. In the Donora episode, for example, one key bit of evidence indicating that the air pollution was severe was the death rate of 17 during the four days of bad air, as compared with a "normal" death rate of 2 for every four days, averaged over the previous 100-year period.

Perhaps the ideal situation is a combination of these two approaches, where researchers can be on hand studying the population while in the midst of the experiment. In the case of the summer camp in Fairview, New Jersey, the researchers set up their monitoring equipment right in the middle of the hustle and bustle of camp activity, during a normal four-week period in late July and early August. The campers went about their regularly scheduled activities—swimming, boating, games—while the researchers recorded what they needed. Every child in the camp was monitored on every day that he or she was there, at least once between the hours of 11 A.M. and 6 P.M. A mobile laboratory was set up which included spirometers to measure breathing capacity and other instruments to continually monitor the ambient air. The children were measured during breaks from their regular activity or during their free periods. On each visit to the mobile laboratory, each child underwent three different examinations: one was a short questionnaire to reveal if he had any symptoms such as hoarseness, runny nose, cough, or watery eyes. Then the researchers measured the peak expiratory flow rate (PEFR) with a specialized spirometer. The third examination involved routine spirometer tests to determine any changes in the capacity of the lungs (FVC), volume (FEV), and other indications of lung well-being.

The results of the daily tests on the children were stored in computer files. While the tests were being run on the children, the air-monitoring devices were also recording such factors as temperature, humidity, air pressure, and the levels of ozone, sulfur dioxide, and other pollutants. Later, in the laboratory, the results of both the spirometer readings on the children and the air quality records were collated and statistically prepared to show lung function as it correlated with atmospheric conditions. The interesting part of the study is that not once during the four weeks of the experiment did the levels of ozone in the air exceed the NAAQS of 0.12 ppm. On most days the ozone levels averaged between 0.08 and 0.10 ppm, higher than ambient ozone levels were 100 years ago but not at levels currently considered dangerous to human health. And that was the whole point of the study, to determine if there is danger to active children even when ozone levels are below the "danger" level.

The researchers found that lung function, in general, did decrease among the children studied during times of increased ozone levels. But they also found a wide range of variability in the results, which they attributed to the wide variability of individual responses to ozone. Some people are more responsive to ozone than others, just as in the plant world some species seem to tolerate high levels of ozone while others succumb to the smallest change. Among the children at the summer camp, the girls appeared to be more sensitive to ozone than the boys, for example, though the reason remains a topic for further study. Also, some of the children (less than a third) in the study showed no decrease in lung capacity or expiratory flow rate while others showed marked changes. In their published conclusion, the researchers conducting this study wrote that the small number of children showing no effects had not altered their general conclusion that children undergoing moderate exercise on hazy, warm summer days experience decreased lung function, and possibly lung damage. They also concluded that the correlations between ozone levels and pulmonary function demonstrated that, for a proportion of the population, ozone levels of 0.08 and 0.10 ppm are still too high for the safety of active children. When this study is compared with the study done earlier in Mendham, New Jersey, where the level of activity of children was much less, it appears that the key factor in ozone responsiveness is the level of activity. The more active the child (or, as other studies have shown, the adult), the more likely it is that he or she will experience decreased lung function. As discussed earlier, this decreased lung function coincides with how ozone reacts on lung tissue. Ironically, it may turn out that exercise is not good for you. While exercising, the exerciser is drawing air in deeper and more often, usually breathing through the mouth and thus bypassing the filtering mechanism of the nasal passages. The ozone in the air then has a chance to react with the more delicate tissues of the bronchioles and the alveoli. While one is at rest, the breathing is generally more shallow, and ozone does not reach the deeper confines of the lung.

As in other recent studies, the researchers questioned the applicability of the National Ambient Air Quality Standards based on average one-hour peak exposures. Current research seems to indicate that longer exposures to lower levels of ozone mirror the real world more precisely, and that federal standards should therefore be based on a more realistic eight-hour average. However, debate continues, and in 1986, in a published review of the NAAQS, the authors, while admitting that there might be something in revising the basis for the standard to an eight-hour average, still maintained that evidence until then did not yet support such a revision. Perhaps that view will change when the evidence from further studies proves overwhelming.

What appears to be clear is that even the levels of ozone found in the ambient air in rural areas, away from the sources of pollution, can affect the lungs of healthy, active children and adults. If no steps are taken, and if ozone levels in the air over the northern hemisphere continue to increase, we can expect more and more increases in the numbers of victims of respiratory illness.

The statistics for these cases may not show up for 20 years or more, but by then it may be too late.

In the meantime, the so-called normal population can take a lesson from those who have suffered chronic respiratory ailments such as asthma and bronchitis. Such patients know that when the air is bad they must reduce their activity. Strenuous activity such as running or playing active sports like basketball should be avoided during days of high ozone levels. The American Lung Association recommends that if you must exercise on days when the ozone level is high, then do so in the early morning hours, when the levels are lower than they will be later in the day. Also, avoid exercising near congested highways. Studies show that a half hour of running in a typically polluted area is the equivalent of inhaling the carbon monoxide from smoking a pack of cigarettes in one day. Also, if you are running in the streets of a crowded urban area, stay at least 30 to 50 feet away from cars, and at traffic lights stay ahead of the exhaust pipes. Obviously, the best course of action during days when pollution is at dangerous levels is to stay indoors and limit your activities. The point is that if the air is bad, then it's best to breathe less of it. And exercising forces the body to breathe more. If the air continues to deteriorate, we may have to accustom ourselves to seeing playgrounds empty on sunny, warm days, and to seeing the streets empty of children during the midday hours. This is a harsh scenario, but it may be necessary to protect human health in the future.

CHAPTER EIGHT

Governmental Policy

Barking up the Wrong Tree

Scientific meetings are not casual affairs. They don't just "happen," in the manner of planning a small dinner party with a few friends, most of which can be taken care of within a week or two of the planned event. Scientific meetings are usually big deals. Many such gatherings are planned in anticipation of a special event, such as the experiment in July and August of 1985, when more than 60 scientists converged on the Amazon basin to study the atmospheric chemistry above it. Well before the actual experiment began, planning was also begun for a meeting in December 1985 in San Francisco, to discuss the preliminary results from the experiment. Nearly 200 scientists from the atmospheric sciences community attended these special meetings, which were part of a larger gathering, the annual meeting of the American Geophysical Union. The scientists heard over 30 presentations describing the most extensive set of measurements ever taken over the Amazon.

In 1986, George Wolff, an atmospheric scientist working for General Motors, in collaboration with John

Hanisch, an administrator for the Environmental Protection Agency (EPA), initiated plans for another specialty conference to take place at the end of 1987. The timing of the meeting was important: December 31, 1987, was the final deadline for all of the states to come up with a plan to meet the clean air requirements as mandated by federal legislation as part of the Clean Air Amendments 10 years earlier. When Wolff and Hanisch began planning the meeting there were still numerous communities around the country that were in violation of EPA standards. Most of these violations were of the ozone standards. And many of the people affected by these violations resided in the Northeast, the region Hanisch administered for the EPA. So the problem to be resolved at the meeting was this: If the air quality in so many areas was still not in compliance of the standards, what could be done? And what should the consequences be if the standards were not met? There were many options, but very few of them were popular.

For ozone, for example, since the automobile was the primary contributor to the problem, some alternatives were considered: mandating construction of new mass-transit systems; imposing the use of car pools on certain roads at certain times of day; and imposing bans on building roads in congested localities until the air quality improved to acceptable levels. These issues were to be discussed at the end-of-the-year meeting. The day of decision was approaching rapidly, and meanwhile there were still too many areas of the country where the air was still too dirty based on the 1977 standards.

So Wolff and Hanisch determined that the end of

1987 was a good time for a meeting to see what could be done to curb ozone pollution. They planned their meeting for the third week in November, to take place in Hartford, Connecticut. Nearly 300 scientists and government officials were invited to attend, from all parts of the country.

But a funny thing happened on the way to the meeting. On the day before the conference was to begin, the day when many of the attendees were preparing to travel to Hartford, Lee Thomas, chief administrator of the EPA, announced he wanted all EPA staff to stay in their offices to listen to an important announcement. In his statement, Thomas announced that the December 31 deadline was to be rescinded. So the purpose of the meeting in Hartford, the urgency, no longer existed. Still, the meeting was held, and most of those who had planned on going went. It was still successful, despite the fact that the reason for its timing was no longer valid.

NO MORE FILTHY AIR

Some good things have happened. The air in the cities of many industrialized countries of the northern hemisphere is cleaner now than it was a few decades ago. In retrospect, a historical look at air pollution always reveals the same distressing pattern: not until there is a crisis in air quality has a government decided to take action. London suffered several episodes of "killer smog" before that government enacted controls on smokestacks, industry emissions, and home heating.

In the United States, not until such catastrophes as the one at Donora and in such cities as Pittsburgh and St. Louis did the federal and local governments take notice. A simple switch to cleaner burning fuel and a simple addition of "scrubbers" to industrial smokestacks solved the problem. But as we've already discussed, the solutions work for only a certain kind of pollution, caused by particulates, smoke, and sulfur dioxide.

The problem in Los Angeles that began to appear after World War II was a very different one. Los Angeles smog was different from the "killer smog" that had plagued London and other cities in the past. In Los Angeles the killer wasn't smog; that is, it wasn't composed of smoke and fog. It was a haze caused by the photochemical reaction between gases and sunlight. Eventually scientists labeled it *photochemical smog,* which may be more accurate but still not exact. It was A. J. Haagen-Smit, of the California Institute of Technology, who in the 1950s led the way by demonstrating that in sunlight, the hydrocarbons and nitrogen oxides emitted from automobiles produce ozone.

The extremely high ozone levels measured in the Los Angeles basin in the 1960s are, for the most part, a thing of the past. During the 1950s and 1960s ozone readings of more than .40 parts per million (ppm) occurred several times a year. Now such readings are rare, almost nonexistent, thanks to the air pollution controls on automobiles that reduce nonmethane hydrocarbon emissions. What such controls have done is to reduce the extent of the bad air; we have cleaned up the worst part of the problem. But the more difficult task of making our air acceptable has yet to be completed.

A PUBLIC POLICY

President Lyndon Johnson's signing of the Clean
Air Act on December 17, 1963, marked a great achievement, a response to our nation's awareness that we
needed national leadership to clean up the air we
breathe. National concern about the environment continued to grow throughout the 1960s, culminating in the
first Earth Day celebration on April 22, 1970. That same
year the Environmental Protection Agency was established. The Clean Air Act amendments of 1970 provided the first substantive effort to quantify the air pollution problem. The 1970 legislation required the newly
appointed EPA administrator to establish National Ambient Air Quality Standards (NAAQS) for a number of
pollutants. The rationale behind the NAAQS was that
the air we breathe should contain sufficiently low concentrations of certain trace gases to ensure protection of
public health. These standards were known as the
primary standards. Secondary standards were also established to "protect overall public welfare," a broad concept that includes natural resources, aesthetics, economics, and the general quality of life.

These amendments required the preparation of
state implementation plans, or SIPs, imposing emission
controls on polluting sources and suggesting other
strategies and methodologies to clean up the air in all
areas within each state. Cleaning up the air to meet all
primary standards established by the EPA was to be
accomplished within three years after each SIP was ap-
proved by the federal agency. The 1970 legislation
stated that each state was to come up with an SIP within

nine months, and the EPA was to approve the SIP within four months of submission. Extensions of two years were granted to states if technologies for reducing specific emissions were not available. In 1976 the final deadline for attaining the primary standards passed by with many areas of the country not in compliance. So Congress amended the Clean Air Act on August 2, 1977, to extend the deadline to 1982. For areas with severe ozone or carbon monoxide problems the deadline was extended even further, to 1987. To make the goals easier to reach, the ozone standard was weakened by increasing the NAAQS from .08 ppm to .12 ppm. The EPA justified this change by citing research showing that ozone at levels of .08 ppm was not harmful to the public. But as discussed in the last chapter, there is no "margin of safety," and the most recent research seems to support a policy of reverting to the more stringent standard of .08 ppm.

The EPA devised other measures to make the NAAQS more easily attainable. One of these was to allow for a one-time violation of the standard each year. So as we enter the last decade of the 20th century the primary ozone standard set by the original EPA in 1970 has been increased by 50 percent, and each area is allowed to violate this newer standard once per year without penalty. Are we any closer to achieving the goal set in 1970? No, especially concerning ozone levels. But all is not bleak. Since the passage of the 1963 legislation, we have managed to reduce the amount of some pollutants in the air. The EPA estimates that lead emissions decreased by nearly 90 percent between 1978 and 1987; sulfur dioxide emissions have gone down by nearly 40 percent; and carbon monoxide by more than 30 percent.

A FALSE SENSE OF PROGRESS

In the late 1970s and early 1980s it actually looked as though ozone levels were decreasing over many of the populated areas of the United States—an interesting paradox considering the subject of this book, that a global increase in tropospheric ozone has been observed in the remote regions of the world during the same period. The EPA could point proudly to the fact that their policy of reducing hydrocarbon emissions had reduced ozone levels in urban areas, just as the research said it should. But there's another side to the story, one that is just now showing itself as we begin the 1990s. Remember that ozone is made from both hydrocarbons and nitrogen oxides, both of which claim as their primary source the automobile. The much-vaunted catalytic converter has done much to reduce the amount of hydrocarbons and carbon monoxide from individual automobiles, but little has been done to reduce the amounts of nitrogen oxide from what the EPA defines as "mobile sources" (cars, trucks, vans, etc.). However, for a few years in the 1970s and 1980s it did look as if the catalytic converter might be enough; that period saw a decline of nitrogen oxide emissions from automobiles. Some of the reasons for this decline could have been the general downward drop in fuel usage resulting from the combination of a rise in prices following the oil embargo in 1973 and the rise of the fuel-efficient cars.

It was at this time that the Japanese car began to make serious inroads into the American automobile market. Because of the high price of gasoline in Japan, the Japanese had always valued fuel efficiency and produced cars that had the best mileage-per-gallon statistics in the world. The competition from Japan spurred Detroit to follow suit, though by the time Detroit began selling fuel-efficient models, the Japanese cars had already cornered a large segment of the market. Still, by the end of the 1970s, with the federally mandated 55-mile-per-hour speed limit and fuel-efficient cars, it looked as if America was at least on the right track.

But then came the 1980s. Fuel prices went down as supplies rose, though prices never approached the pre-1973 levels. Gasoline usage began to rise steadily. Sure, American cars were much more fuel-efficient than they had been, but there were also more cars on the road. The ultimate effect has been that our fuel consumption now is greater than it was in the 1970s. With more cars using more gasoline, the amount of the nitrogen oxides emitted into the atmosphere has risen to greater levels than before the introduction of the catalytic converter. And perhaps the most ominous sign is that automobile advertising again emphasizes speed over efficiency. Meanwhile, the federally mandated speed limit has been raised back up to 65 miles per hour.

DEPRESSING STATISTICS

All of these circumstances appeared to catch up with us during the summer of 1988, one of the hottest on record. The season started out warm, and by mid-June a large stagnant air mass had settled over the eastern United States, where it remained most of the summer. With the stagnant air mass pumping hot, humid tropical air up from the Gulf of Mexico and the South Atlantic, and with little movement of air systems to bring relief, ozone concentrations in many parts of the United States reached record levels, and these levels remained high sometimes for days on end. Even in Acadia National Park on the coast of Maine, park officials had to warn visitors to "take it easy," as ozone levels reached .14 ppm, the first time the park had ever been exposed to such high concentrations.

In the Shenandoah Mountains of southwestern Virginia, the EPA has set up several monitoring stations to measure various trace gases. During the summer of 1988 measurements at these stations violated EPA standards twice, the first time that concentrations so high had been measured since the stations had been established nearly a decade earlier. Not only was the air so bad that even rural areas like the mountains of Virginia and the coast of Maine experienced record levels of ozone, but the cities also suffered. Detroit, for example, violated the ozone standards 16 times in the summer of 1988, compared to 3 times in all of 1987. Philadelphia exceeded the standards 42 times, compared with 31 for 1987, and cities such as Charlotte and Cleveland had almost twice as many violations in 1988 as in the previous year.

Virginia and Maine are good examples of how the bad air problem has changed since the early 1980s. Both states are predominantly rural, both states rely on tourism as an economic base, and both have little heavy industry to pollute the air. Being in the business of tourism, both states view their air and water as valuable economic resources. People do not travel to Maine to breathe in the same noxious air they breathe in New York City, nor to swim in a lake which they have to share with sewage. The people and governments of both states view clean air and water as more than just cosmetic window-dressing to making a living. Clean air and water *are* the living. In Maine, the state standard for ozone remained at .08 ppm even after the national standard was lowered. Virginia has an extensive system of 19 monitoring stations to measure ozone levels, spread across the state. Maine has 10. During the summer of 1988 there were 110 violations of the ozone standard in Virginia. The previous high had occurred during the summer of 1983, when 65 violations were measured. Another state on the eastern seaboard, Connecticut, also had record-high ozone levels and record occurrences of violations, but in Connecticut's case the problem was its proximity to New York City. New York City, during the summer of 1988, was in violation of federal ozone standards 40 percent of the time. Thus New York City's problem became Connecticut's problem and eventually became the problem of the visitors to Acadia 300 miles up the coast.

Such occurrences in nonindustrial states like Virginia and Maine underscore one of the most distressing problems in ozone regulation. The polluted air over those states, for the most part, didn't originate there. The ozone levels experienced in Maine originated to the south and southwest, and the ozone in Virginia may have originated in the refineries and industrialized areas of the Texas and Louisiana Gulf Coast. Some studies have shown that some of the pollution measured over southeastern Virginia originated in the New York–Boston corridor. From a regulatory point of view, there is little that can be done by officials in Virginia or Maine or Connecticut to decrease ozone levels in their states. And therein lies the problem that has been the focus of this book: Ozone levels everywhere in the northern hemisphere have increased considerably over the last several decades, and therefore air entering a remote region doesn't have to be made that much dirtier to be in violation of EPA standards. Virginia's plight, for example, highlights the complex nature of the interaction of spatial scales. Even though air quality on the local urban scale may arguably be better than it was years ago, the dirtier air on a global scale is being found in such nonindustrialized states as Virginia. Granted that the summer of 1988 was not a "normal" summer, but if the climate is changing because of increases in carbon dioxide concentrations as the climate modelers predict, then the summer of 1988 may be only the first sign of what will be a "normal" summer of the future.

THE PARADOX OF SCALES

If the air in our cities is cleaner, then why is the air away from our cities dirtier? Such is the paradox, a paradox promoted by the EPA policies of the past. A policy to reduce ozone formation in the Los Angeles area focused primarily on reducing hydrocarbon emissions, even though hydrocarbons are only one of the three components necessary to generate ozone (the other two being nitrogen oxides and sunlight). Since sunlight cannot be regulated, the only other component remaining is nitrogen oxides, and it is these that have escaped the stringent regulation imposed on hydrocarbons. The technology exists to reduce nitrogen oxides, but it is somewhat more sophisticated than the technology used to reduce hydrocarbons or carbon dioxide coming out of the automobile's tail pipe, or sulfur oxides coming out of smokestacks. A catalytic converter exists which can transform nitrogen oxides into harmless nitrogen molecules, the most abundant gas in our atmosphere. But it is more expensive than the existing catalytic converter. The auto industry has logically argued that the catalytic converters currently required in California have reduced the smog problem in the Los Angeles basin, and that therefore the auto industry has done its part to solve the problem. Basically, the debate boils down to a cost-vs.-benefits battle, with the auto industry claiming that the less expensive hydrocarbon catalytic converter has done the job without the added expense of the nitrogen oxide converter. On their side, research has shown that by reducing nitrogen oxide emissions we may end up increasing ozone levels in the urban centers. So, by reducing hydrocarbon emissions in the urban centers we have inadvertently dirtied the air outside the urban centers; hence the widespread elevated ozone concentrations during the summer months of stagnant air. How can this be?

To understand the paradox, it helps to look at the specifics of the chemical reactions that take place in the atmosphere. Ozone and nitrogen oxide react rapidly to create nitrogen dioxide (NO_2) plus oxygen (O_2). If we stopped there, we can see how adding nitrogen oxide to the atmosphere would reduce the ozone, since it would break apart the ozone molecule (O_3) and transform it into oxygen. But most of the nitrogen dioxide created by this reaction breaks apart when exposed to visible light and ends up giving back nitric oxide (NO) and an oxygen atom. These immediately re-form ozone by combining an oxygen atom with an oxygen molecule (O_2).

These three reactions cause a situation which photochemists call the *photostationary state*. It means there is a certain ratio between nitric oxide (NO), nitrogen dioxide (NO_2), and ozone (O_3) that is determined by the rates of the reactions, by the actual amount of the trace gases present, and by how efficiently nitrogen dioxide is broken down by sunlight. Since nearly all the emissions of nitrogen oxides (i.e., NO_x, which is the sum of both nitric oxide and nitrogen dioxide) is in the form of NO, the photostationary state tends to reduce ozone near the source of the nitric oxide, the primary pollutant.

The photostationary state is an important numerical relationship. It makes simulation of the processes that produce smog easier to calculate on the computer. But when scientists take direct measurements of each of the products involved they find that the photostationary state does not precisely exist. Generally, the measurements of the gases involved (NO, NO_2, and O_3) and of the photolysis rate of NO_2 compute a ratio a few percentage points different from what the theory says it should be. It is this imbalance between what the photostationary state should be and what it actually is as measured in the atmosphere that drives the formation of ozone.

What happens is that the initial reduction of ozone near the source of nitrogen oxide also results in an increase in nitrogen dioxide, a chemical that is generally not found in the atmosphere at levels deemed by the EPA to be harmful to the environment. Although it is not considered a pollutant, the nitrogen dioxide has another function which contributes to the pollution problem: it serves as a hidden reservoir for ozone. By reacting with other gases in the atmosphere, nitrogen dioxide is transformed into other chemicals that are not part of the photostationary state. With all the NO_2 present in the urban atmosphere because of local NO emissions, a small fraction of NO_2 is not broken down by sunlight. Instead, it reacts with something else, often a fragment from the hydrocarbons also emitted from auto exhaust. The chemical species created by such reactions, of which there are dozens, are called *organic nitrates*. Organic nitrates are any gases that contain both nitrogen and carbon. Most of these compounds are present in the atmosphere at such low concentrations that they cause no direct harm to the environment. But by reacting with NO_2 they can indirectly affect the production of ozone.

Another possibility is that the NO_2 is converted to nitric acid, one of the prime components of acid rain. So it is easy to see how the abundance of nitrogen oxides in the cities may reduce ozone levels there while the ozone that has transformed it to nitrogen dioxide can be carried into outlying areas in the form of organic nitrates and nitric acid.

When these organic nitrates and nitric acid reach the remote regions of the atmosphere, they undergo reactions which eventually release the same nitrogen dioxide that went into their formation in the first place. However, these reactions are generally slow, so the organic nitrates and nitric acid can be transported relatively long distances before giving back their nitrogen dioxide to the atmosphere. Once this happens, the nitrogen dioxide is broken apart by sunlight, and the result is a rapid formation of ozone. In other words, this ozone formed over remote areas is the same ozone that would have been created in the urban areas closer to the source of the nitrogen oxide; instead, it has been transported away from its source.

What this all means is that now there is a fresh source of nitrogen oxides (both nitric oxide and nitrogen dioxide, as the photostationary state constantly tries to reestablish itself) in the rural atmosphere that, paradoxically, got there because at its source it was "left over" from urban pollution. So now the rural areas are reaping the "benefit" of the urban areas' "cleaner" air in terms of ozone pollution, the leftovers from an abundant feast. But in order to make ozone, there needs to be a third ingredient besides sunlight and nitrogen oxides. In the rural atmosphere, this third ingredient is either one of two common, ubiquitous trace gases: methane or carbon monoxide. So despite the lower levels of ozone in our large urban areas, there is still too much nitrogen oxide being emitted, too many cars and internal combustion engines. It is the raw materials being produced in our cities that are the cause of the rise of ozone levels in our rural areas, and that result in the widespread high ozone concentrations found in Figure 25.

"POLLUTING" TREES: PROOF OF A FAILING POLICY

In 1988 and 1989, two papers appeared, one in *Science* and the other in the *Journal of the Air Pollution Control Association,* written by a group of scientists at the Georgia Institute of Technology. Led by Bill Chameides, with Ronald Lindsay and Jennifer Richardson, the scientists set out to determine whether or not the reduction of hydrocarbons in the Atlanta area would result in

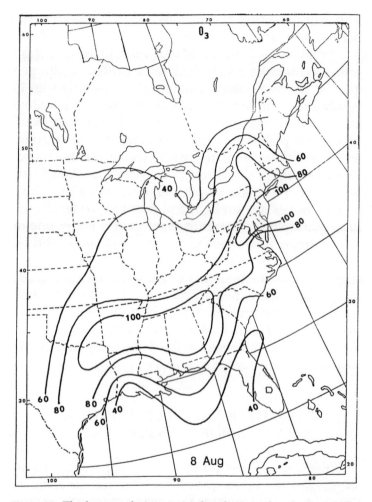

Figure 25. The large-scale (synoptic) distribution of surface ozone for August 8, 1980. This distribution has been derived from an analysis of more than 200 EPA monitoring stations over the eastern United States. August 8 was the last day of a meteorological situation that resulted in the presence of the most widespread air stagnation over

a decrease in ozone levels. The EPA had estimated that hydrocarbon emissions in the area had decreased by nearly 40 percent between 1979 and 1985. According to the prevailing theory at the time, such a reduction should have affected ozone levels, reducing them by about 20 percent. But such was not the case. After analyzing the data, Chameides and his group could find no reduction in ozone levels at all. In fact, they concluded that ozone levels either had remained the same or had increased slightly. Why?

Trees. Their studies showed that the amount of the hydrocarbons released from manufactured sources in the Atlanta area was approximately equal to the amount of the hydrocarbons released by trees growing in the region. Trees release two types of hydrocarbons: isoprene, comprised of 5 carbon atoms, and pinenes, comprised of 10 or more carbon atoms. The pinenes, being heavier molecules, often settle on small particles present in the atmosphere. They are responsible for the haze often seen in forested regions, especially the famous Smoky Mountains in southern Appalachia. Pinenes are also responsible for the piquant pine scent of the pine forests.

Isoprene, on the other hand, can eventually contribute to the formation of ozone. It was this fact that made the headlines when Chameides's study was re-

---

the southeastern United States during the 1980 summer. The isopleths are in units of parts per billion, and no stations in or downwind of urban areas were used in the analysis. This particular event illustrates how widespread ozone pollution can be. This pollution is especially worrisome in the southeastern United States, where many forests may be susceptible to ozone damage.

leased: "Trees pollute." Actually, trees don't pollute. Many ecologists believe that the isoprene emitted by trees is a reaction to ozone levels and may be the trees' defense against ozone damage. The isoprene from trees reacts with ozone and neutralizes ozone's potential effects on plant cells before it enters through the stomata. But isoprene also reacts with the ubiquitous hydroxyl radical. In a heavily forested area, most of the isoprene reacts with the hydroxyl radical, and any left over reacts secondarily with ozone. Without any nitrogen oxides, the fragments of the isoprene molecules that are formed eventually go through a convoluted series of chemical reactions which transform them into organic acids or organic peroxides. These trace gases never build up to harmful levels in the atmosphere and are usually washed out by rain.

In a natural forest environment, the concentrations of nitrogen oxides are usually very low, so the hydrocarbon given off by trees is harmlessly transformed into substances other than ozone. But with levels of nitrogen oxide high, the isoprene fragments furnish the RO_2 radicals discussed in Chapter 3, which are necessary for the rapid formation of ozone. It doesn't matter if these RO_2 radicals came from an automobile exhaust or a tree. What's important is their presence, along with sunlight, in the atmosphere. This is what the Georgia Tech researchers concluded from their study. They reached this conclusion by conducting an inventory of how many manufactured hydrocarbons were released in the Atlanta area and how many were released naturally from trees. They used land-use statistics and satellite imagery from LAND-SAT to compile figures to quantify these emissions. What they discovered was that hydrocarbon emissions from vegetation and the amounts released by autos and other manufactured sources (based on EPA figures) were approximately the same for the year 1978, the baseline year for their study. Between 1978 and 1985, more than $700 million was spent by the government to reduce manufactured hydrocarbon emissions by 37 percent in the 11-county area that comprises greater Atlanta. From satellite data the Georgia Tech scientists were able to determine that forested areas had been reduced slightly because of urban sprawl during the study time. Again, using EPA data, they determined that nitrogen oxide emissions had increased by 4 percent.

The tricky part of the study was to determine exactly what effect the $700-million investment had had on ozone formation. The problem was that all the ground-based EPA monitoring stations in the area showed an increase in ozone levels during the time period. What was not obvious was just how much of this increase was due to widespread increases in tropospheric ozone throughout the hemisphere and how much was due to local production within the Atlanta metropolitan area. To quantify just how much ozone production had been locally produced, the researchers examined data from the outskirts of the 11-county area and screened ozone measurements by using only data from particular wind directions. For example, they compared ozone measurements from one station northwest of the city with those from one southeast of the city. During the first year studied, 1978, when the wind was from the northwest the ozone concentration at the southeast station (downwind of Atlanta) was higher by about .03 parts per million. Likewise, when the wind direction was from the southeast, the station located northwest of the city measured ozone concentrations about .03 ppm higher. Thus the researchers concluded that pollution from the Atlanta metro area accounted for an ozone production rate of about .03 ppm.

Thus if reduction of hydrocarbon emissions had truly been responsible for reducing ozone formation in the atmosphere, then the researchers should have observed a smaller difference in ozone concentrations at the two sites observed. But by 1987, the last year in which the data were studied, the difference between the two stations, when screened by wind direction, had increased to more than .035 ppm. So the trend in ozone production in the Atlanta area was *up* by 3 percent per year, despite the success in reducing hydrocarbon emissions. The Georgia Tech researchers concluded that hydrocarbon emissions from the trees in the area were sufficient to generate ozone effectively despite controls put on manufactured hydrocarbon emissions. By using a photochemical model, the researchers demonstrated that the most effective way to reduce the ozone formation potential in the Atlanta region was not by reducing hydrocarbon emissions but by reducing nitric oxide emissions. Cutting down trees would have the same effect as cutting hydrocarbon emissions from automobiles—a very small impact on the large-scale production of ozone.

NEW STRATEGIES FOR OZONE CONTROL

We've increased our knowledge of how ozone is formed in the atmosphere considerably since the early days. When Haagen-Smit began studying the problem of photochemical smog in the 1950s, there wasn't a hint that the acute problem faced almost exclusively by residents of Los Angeles extended much beyond the hills of southern California. But we now know that the problems that plagued Californians are very similar to the ones that discourage visitors to Maine's Acadia National Park, on the other side of the continent. And we now realize that many of the air pollution controls put on automobiles to reduce hydrocarbon emissions will not solve the entire problem.

A summary of the scientific gains made since research began in the 1970s reveals this striking truth: *Emissions of nitrogen oxides, in addition to hydrocarbon emissions, must be controlled to solve our national ozone problem.* The technology already exists. In California, controls on nitrogen oxide emissions from automobiles are already in effect. If President Bush wants to be known as the "environmental president," as he claimed during his 1988 campaign, then he must see to it that nitrogen oxide emissions are controlled.

We began this chapter with the story of how scientific meetings are often planned to coordinate with anticipated milestones. Perhaps the future will see a scientific gathering called to discuss the effects of new air-pollution technology on ambient air quality. Hopefully, the results by then will be beneficial.

South of the Border
The Tropical Ozone Dilemma

The National Center for Atmospheric Research (NCAR) in Boulder, Colorado, is a stunning piece of architecture located on Table Mesa, a plateau situated just before the foothills of the Rocky Mountains (see Figure 26). Sitting 600 feet above the city of Boulder, the NCAR building attracts as many visitors for its architecture as it does for its research. The building is considered by many to be the crowning achievement of its illustrious architect, I. M. Pei. With its twin towers, the six-story building blends in remarkably with the towering mountains in the background. But many visitors to NCAR are professional scientists, there not to gaze at the beauty but to roll up their sleeves and participate in one of the premier centers funded by the National Science Foundation. NCAR is operated by a consortium of universities in the United States and Canada which have curricula in either atmospheric science or oceanography. Because of its special place, both geographically and within the university milieu, it is a popular location for professors from around the country eager to spend their sabbatical leaves doing research.

In 1977, one of those eager professors was George Dawson from the University of Arizona. He specialized in measuring various trace gases in the background atmosphere. Before coming to NCAR Dawson had arranged to work with Paul Crutzen to study the emissions that originate in cattle feedlots. Dawson's area of expertise was atmospheric ammonia, a primary source of which, it was believed, was animal urine. Dawson had developed a special instrument that could measure concentrations of atmospheric ammonia down to less than one part per billion. Our knowledge of ammonia is crucial because it is the most abundant alkaline trace gas in the atmosphere. Unlike some other trace gases, ammonia turns basic (the opposite of acidic) when it dissolves in rainwater. Trace gases such as sulfur dioxide and nitric oxide, if they eventually undergo chemical reactions to become part of rainwater, help to increase rainwater's acidity, thus making acid rain. Ammonia is a noteworthy exception to the acid rain problem: increasing concentrations of ammonia in the atmosphere actually decrease the acidity of rainfall.

Dawson had planned to place his instrument inside and downwind of the commercial cattle feedlots that dot the plains of northeastern Colorado, not far from Boulder. His plan was to collect as many samples as possible to determine the atmospheric concentration of ammonia under a variety of atmospheric conditions,

---

Figure 26. A view of the National Center for Atmospheric Research (NCAR) in Boulder, Colorado. NCAR is one of the leading centers in the world for research in atmospheric chemistry. (Photograph courtesy of the National Center for Atmospheric Research of the National Science Foundation.)

then to quantify the magnitude of this ammonia source. Perhaps spending one's sabbatical hanging around feedlots doesn't seem very glamorous, but someone had to do it, and it was, after all, Dawson's specialty.

But Paul Crutzen had another idea for Dawson to pursue while he hung out with the cows. As discussed in Chapter 3, Crutzen's primary interest in the late 1970s was determining the quantity of trace gases released to the atmosphere as a result of biomass burning. In his studies Crutzen discovered that one of the primary fuels used for domestic cooking in India is cow dung. Because of India's large human population and large number of cattle (which are protected by religious and legal sanctions), Crutzen estimated that significant amounts of methane and other trace gases result from the burning of this fuel and that this source of methane constitutes a substantial portion of its global sources. But Crutzen was missing one important factor for his calculations: he had to know how much methane was released when the cow "paddies" were burned.

So Crutzen asked his visitor, Dawson, to do him a big favor. Would George mind terribly much collecting a few samples for him? And one more thing, they had to be fresh. Fresh cow dung. Crutzen had read that women in India gather their fuel immediately after it is discharged from the animal. So to simulate a process happening halfway around the world as accurately as possible, Crutzen emphasized to Dawson, the dung had to be fresh. Aged doo-doo wouldn't do. And since in India the fresh cow chips are baked in the sun for several days before burning, Crutzen arranged for the paddies to be placed on the roof of I. M. Pei's masterwork to dry. The roofs of the twin towers of the building are a favorite haunt for weary researchers, who sit out there and clear their minds with the grand view of the mountains. But for the few months during George Dawson's visit, the scientists had to be careful where they stepped.

A WORLDWIDE CHALLENGE

The above example illustrates how sources of pollution created by humans (cows' excrement wouldn't be burned, except for civilization) are somewhat different in the tropics than at mid latitudes, even though the end result at both low and high latitudes is the same: the air becomes dirtier. The industrialized nations of the world are able to meet the challenge of reducing pollution by implementing existing technology to reduce emissions. It costs money, but the benefits justify the cost. In the less developed nations, it's another story. We are faced with the dilemma of trying to place pollution controls on less developed countries that are trying to achieve a better standard of living. It's as if the developed nations have just finished a grand feast while the undeveloped nations waited in the wings, and now we're going to tell them there is no more food. Easy to say with a full stomach. Such densely populated cities as Mexico City, São Paulo, and Lagos are known for their bad air, and yet pollution controls in those cities are virtually nonexistent. Furthermore, regarding ozone, the potential for the generation of photochemical smog is even worse because nearly every fast-growing major city in the world happens to be situated at a low latitude where the energy of the sun can "cook" the pollution soup from the city with greater efficiency.

Figure 27 shows the projected population of 12 of the largest metropolitan areas by the year 2000. Of these cities, only three—New York, Tokyo, and Seoul—are located in developed, industrialized countries in the middle latitudes. The rest are in the tropics or semi-tropics, in underdeveloped or developing nations. This is not surprising, since the population, in general, in the tropical and subtropical latitudes is growing three times faster than the population in northern mid latitudes. This growth is taking place not only in the cities of the tropics, but in the rural areas also. To support this in-creased growth, members of these growing populations are relying more and more on the agricultural practices that have traditionally used biomass burning to feed them. Even though such practices have been used for centuries, they will become even more widespread and prevalent as the population increases and there is more pressure for more food. Satellite measurements show that already there is three to four times more ozone present over tropical areas influenced by widespread biomass burning than over tropical areas not impacted by such practices.

The situation in Brazil poses a particular problem as the nation tries to cope with its growth. In 1980 Brazil made the decision to become as energy-independent as possible. To accomplish this, the government decided to manufacture automobiles that would use alcohol as fuel rather than gasoline. The reason was simple: gasoline had to be imported while alcohol could be produced locally from sugar cane. Thus, during the 1980s the pro-

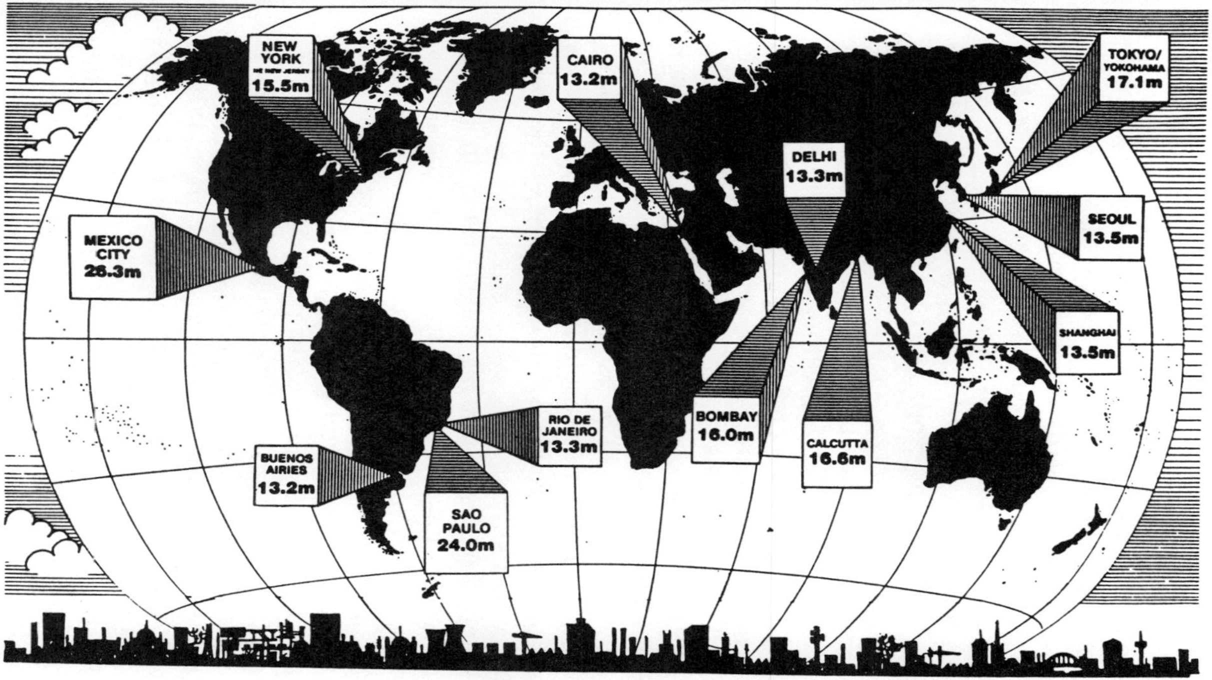
NEW YORK
15.5m
CAIRO
13.2m
TOKYO/ YOKOHAMA
17.1m
DELHI
13.3m
SEOUL
13.5m
MEXICO CITY
26.3m
SHANGHAI
13.5m
RIO DE JANEIRO
13.3m
BOMBAY
16.0m
CALCUTTA
16.6m
BUENOS AIRIES
13.2m
SAO PAULO
24.0m

Figure 27. The projected populations of the 12 most populous cities in the world by the year 2000. Nine of these cities are in developing countries located in tropical or subtropical climates. (Figure from "State of World Population Report" from the United Nations Fund for Population Activities.)

duction of sugar cane increased several-fold while gasoline imports dropped. From an economic view the program has been successful, the Brazilians see less of their
wealth going out of the country, and at the same time
the sugar cane industry has flourished.

But there is a dark side. If sugar cane is to be harvested efficiently, the crops must be burned before the
workers go out to harvest the stalks. Typically, a field of
sugar cane is harvested by an itinerant family with four
or more children. The day before the family goes into
the field to harvest, the area is set afire to burn the thick
undergrowth that grows along with the cane. Only
about 10 or 15 percent of the sugar cane is lost in the
fire, but that loss is overcome by the savings in labor,
because after the field is burned the family has to cut
only the stalks that will be used to make sugar and not
all the other growth that would have been there had the
fields not been burned. But the burning of the fields
produces bad air.

Volker Kirchhoff and his colleagues from Brazil's
Space Agency have placed instruments in the field to
measure carbon monoxide and ozone just outside of
Cuiaba, a city in south central Brazil located near a region in which much sugar cane is grown and harvested.
The results of this study were published in the May
1989 issue of *Geophysical Research Letters*. The bars in
Figure 28 show the monthly average concentrations between August 1987 and October 1988 at Cuiaba. What is
obvious from these graphs is how much pollution is
generated from the burning of the sugar cane fields during the August–October period. An analysis of satellite
data in the same paper shows that the number of fires
during these months may exceed 20,000 per week. Dur-

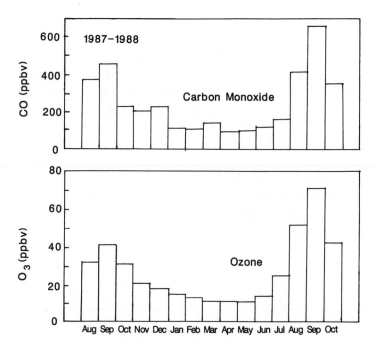

Figure 28. The monthly average concentrations of (ppbv–parts per billion, by volume) of carbon monoxide and ozone measured at a location in central Brazil. The period of these observations covers August 1987–October 1988 and includes parts of the two "dry seasons" when widespread agricultural burning is most prevalent. This biomass burning causes the very high concentrations measured during September of each year. The amount of burning in this part of Brazil has gone up dramatically since 1980 as a result of the increased production of sugar cane. Alcohol is distilled from the sugar cane and is subsequently used as fuel for most of Brazil's automobiles.

ing the entire dry season, lasting from mid-July to mid-October, hundreds of fires are observable by satellites on any given day.

In September 1988, the average ozone concentration in Cuiaba was more than .07 parts per million (70 parts per billion). In comparison, in Fairfax, Virginia, a suburb of Washington, D.C., the National Ambient Air Quality Standard of .12 ppm (parts per million) was violated 15 days during the months of June and July 1988. The monthly average ozone concentrations during those two months was .05 ppm, or nearly one-third lower than the average monthly concentration at Cuiaba during September of that year. In this particular case, and in similar cases, it is safe to say that the ozone pollution in south central Brazil during its high ozone season is worse than the ozone pollution outside one of the major cities in the United States during the peak of its high ozone season.

DEFORESTATION

The pollution generated from the burning of sugar cane fields takes place in a region of Brazil that under "normal" circumstances would be characterized by vegetation common to savannah regions. If the area had not been turned over to the cultivation of sugar cane, the natural flora would be tall grasses and bushland-type plants. The Amazon basin of South America accounts for an estimated 66 percent of the world's remaining rain forests, approximately 2 million square miles (more than half the size of the United States). Most of the Amazon's rain forests lie within the borders of Brazil. The rain forest region lies to the north and west of the sites where Kirchhoff made his carbon monoxide and ozone measurements. In the early 1980s, the international scientific community expressed concern regarding the clearing of a considerable portion of the Brazilian rain forest for economic development (such as sugar cane fields) and resettlement of thousands of Brazilians from overcrowded cities to the south. For example, the population of the state of Rondônia, in the central western part of the country, was growing at the staggering rate of 15 percent per year.[1]

The extent and rate of forest clearing in the Amazon basin are subjects of controversy, although satellite data have helped to minimize the uncertainty. The lack of any systematic data on the subject has hampered discussion of what many scientists consider an ecological crisis with serious biological, climatic, and political ramifications. When coming up with estimates, scientists have speculated that between 10,000 and 100,000 square kilometers of rain forest are being cleared each year. Most of the estimates have been based on surveys taken by state officials, but because of the vastness of the region and the remoteness of the rain forest, their techniques are not considered very accurate. But recent satellite data show quite dramatically how the landscape of Rondônia has been altered. Based on such data, scientists estimate that about 20 percent of the rain forests in that state have already been converted to agricultural use. Unfortunately, much of the land cannot sustain agriculture, and instead of providing food for export as planned, the state imports many of its staples.

[1]*National Geographic*, December 1988, pp. 772–799.

Rain forests are important because they play an integral part in the global carbon cycle. Although historically much attention has been focused on the increase in carbon dioxide resulting from burning fossil fuels, scientists have also identified the biosphere (the portion of the atmosphere that interacts with living matter) as a key factor. Since tropical rain forests contain an estimated 40 percent of the world's plant mass carbon, and tropical seasonal forests another 14 percent, the future of the world carbon dioxide problem and how it relates to global climate could be affected by the fate of the tropical rain forests.

What complicates the issue is that even after the forest is cleared and subsequently burned, a substantial fraction of the biomass that has been cleared is not burned completely. Much of what isn't completely burned remains as charcoal. Without the formation of charcoal the burning of tropical forests would add to the greenhouse effect in two different, and important, ways. First, the burning of the biomass would put carbon dioxide directly into the atmosphere; second, the smaller amount of forest remaining would result in less carbon dioxide for photosynthesis. The formation of charcoal creates a short circuit in the global carbon cycle which can result in a net *loss* of carbon dioxide, rather than an increase.

This contradictory hypothesis was presented by Wolfgang Seiler and Paul Crutzen and published in the scientific journal *Climate Change* in 1980. What Seiler and Crutzen suggested, for the first time, was that the carbon that had been tied up as organic matter in a living forest was, as a result of burning, now tied up as elemental carbon in the charcoal and thus does not find its way into the atmosphere as carbon dioxide. They also pointed out that much of the forest that is burned during a year eventually becomes reforested land, because the cleared and cultivated land that was once forest quickly loses its nutrients and cannot profitably support the growth of crops. In other words, in actual practice the forests are cleared and burned to grow crops, but soon this cropland loses its fertility and begins to revert back to forest. That has been the cycle of slash-and-burn agriculture for eons.

Despite what was clear to Seiler and Crutzen, there were uncertainties: exactly how much of the burned forest was reforested and exactly how much remained behind as charcoal. Because of the uncertainty of these statistics, the two scientists concluded that the burning of tropical forests could result in either a small gain or a small loss in carbon dioxide. This hypothesis was considerably different from those in other studies published prior to 1980 about the effect of deforestation on the global carbon balance.

How can such discrepancies exist? The main problem is the lack of credible data. The data available are rudimentary, even primitive. The Amazon rain forests are remote, rugged, and difficult to get to. Compiling data on such things as forest biomass and carbon content, growth rates of second growth, and just how much of the burned material actually gets transformed into carbon dioxide is what is needed but it's also what is so difficult. Research to fill in the gaps of our knowledge of rain forests should be a top priority, according to Phillip Fearnside of Brazil's National Institute for Research in the Amazon (INPA) in Manaus. Fearnside is one of the world's leading experts on the global impact of the deforestation of the Amazon rain forest. In an article published in *Change in the Amazon Basin, Volume I: Man's Impact on Forests and Rivers*, Fearnside urged that a vast number of resources be invested to fill the knowledge gap about rain forests in general and the Amazon rain forests in particular, especially since the latter represents about 20 percent of the planet's carbon reservoir in living biomass.

THE SILENT THREAT: OZONE AND THE RAIN FORESTS

So we know from Volker Kirchhoff's carbon monoxide and ozone measurements that pollution is a problem even in remote Amazonia. And we also know that the vanishing of the rain forests is an important part of the global carbon cycle and future global warming. But what happens to all that ozone that Kirchhoff measured? Do high ozone concentrations over the cultivated areas mean that the rain forests, in close proximity, are affected, too? Scientists like Barry Rock have found that forests in the United States and Europe are affected by ozone pollution, yet no one has yet looked at the impact of high ozone concentrations on Brazil's rain forests. If the effect on our New England forests also applies to the tropics, then the ozone produced by widespread burning of fields may have an effect on the growth of the jungle. The effect may be far less dramatic than the felling of the large timbers, but it may have a considerable impact on the forest over a longer period of time.

If the growth rate of the forests is adversely affected by the higher concentrations of tropospheric ozone, then the size of the forests will accordingly be reduced. The effect of such smaller forests will be a feedback to the earth climate system by soaking up less carbon dioxide (fewer trees mean less carbon dioxide absorbed) from the atmosphere. The resulting smaller sink will allow carbon dioxide to build up at an even faster rate than if the forest growth rate had not been reduced and had not thus amplified the warming of the atmosphere because of the greenhouse effect. Such a complex synergistic scenario has not been tested in global climate models, but should be considered as another possible consequence of biomass burning in the tropics.

We must also consider the direct effect of the increased amount of tropospheric ozone on the earth's climate. As discussed in Chapter 5, an increase in tropospheric ozone will add directly to the greenhouse effect. The simple climate models used to date all indicate that the impact of tropospheric ozone on climate is most pronounced when ozone levels are raised over the tropics, especially at higher altitudes in the tropical troposphere. Thus, increases in tropospheric ozone at low latitudes (within the tropics) would have a greater impact than the same increased ozone levels at the higher latitudes because the process of vertical mixing by cumulus clouds is so much more efficient in the tropics than in the temperate zones. In other words, the ozone produced by the burning of fields and other biological matter in the tropics is carried to higher altitudes more rapidly, where it can more effectively contribute to the greenhouse effect.

But there's another aspect to the relationship between tropospheric ozone and rain forests: the rate of deposition (loss) of ozone to the surface is dependent on the type of surface the ozone comes in contact with. Scientific studies have shown that the most efficient type of terrain for this process is forests. In other words, forests soak up more ozone than any other terrain. So if the extent of the tropical rain forest is reduced, then one important factor in the global ozone budget will become smaller, and the accumulation rate of ozone in the tropical troposphere will become even greater—once again providing a positive feedback to climatic warming.

AFRICA: THE NEXT FRONTIER IN ATMOSPHERIC CHEMISTRY

When we discuss air pollution, the first thought that comes to mind is the dirty air produced by the industrial cities of the world. We've already seen, however, that huge amounts of pollutants are also emanating from Brazil and other tropical countries, and that these emissions are also contributing significantly to the global pollution problem. But the only reason we know about the pollution in Brazil is that we have a means of measuring trace gases there. If we had measurements of pollution from all over the world, what kind of spatial patterns would they show?

The best way of measuring trace gases over such a large area as the whole world is with a satellite. At the NASA Langley Research Center in Hampton, Virginia, Hank Reichle and his co-workers have developed an instrument that while traveling in space can measure the amount of carbon monoxide present in the middle of the troposphere (between approximately 15,000 and 30,000 feet). This instrument, called MAPS (measurement of air pollution from satellites), flew aboard the space shuttle in November 1981 and again in October 1984, and three more flights are planned between 1992 and 1995. Reichle and his co-workers published two papers in the *Journal of Geophysical Research*, one in 1986 and another in 1990. The published results surprised the scientific community, for they showed that the highest concentrations of carbon monoxide measured were located over Africa.

These findings of high carbon monoxide levels over Africa caused quite a stir in scientific circles. There were questions raised. Because of some technical difficulties during the mission, the MAPS instrument obtained data for only about nine hours, and because of the orbit the shuttle was locked into, the measurements were obtained only between 38 degrees north and 38 degrees south. However, even if there had been high carbon monoxide concentrations over northern midlatitudes (beyond the range of the instrument), the orbit of the shuttle would not have allowed for the pollution to be recorded unless the carbon monoxide had been transported southward, an unlikely scenario because of the prevailing winds. To complicate the interpretation of the measurements further, the sensor on MAPS does not actually measure carbon monoxide at ground level; it is sensitive only to carbon monoxide concentrations that have been lifted to an altitude of 10,000 feet or more. Such rapid vertical transport is more prevalent in the tropics than at mid latitudes. Therefore, carbon monoxide is more likely to be found at high enough altitudes for precise measurement in the tropics, where there are more tall cumulus clouds to carry the pollution rapidly from the surface, than at higher latitudes.

When the MAPS instrument was flown again in October 1984, Katherine Sullivan, one of the astronauts aboard the flight, took photos of thousands of fires she observed along the east coast of southern Africa as the space shuttle flew an orbit that covered the earth between 53 degrees north and 53 degrees south. The 1984 flight took carbon monoxide measurements for six days. Once again, after the data were analyzed, Reichle and his co-workers noted that the highest concentrations of carbon monoxide occurred over central and southern Africa. They also found high concentrations over central Brazil and eastern Asia, and near the coast of Europe. Although they did not measure large amounts of carbon monoxide over the United States, a persistent cloud cover limited the number of data they could obtain since the MAPS cannot measure through clouds. What is obvious from these two space flights with MAPS is that there is a significant air-pollution problem in the tropical areas of the world.

In another research project at Langley Research Center, two sets of satellite data are being used to determine the distribution of ozone in the troposphere. Since the 1970s, several satellites that measure ozone in the atmosphere have been orbiting the earth. When they were launched, the scientific intent of these instruments was to provide a means of studying the amount of ozone in the stratosphere, since approximately 90 percent of the earth's ozone is located there. One instrument, launched aboard the *Nimbus 7* satellite in October of 1978, is called TOMS (total-ozone-mapping spectrometer). Although designed to last for only 2 years, TOMS was still making measurements more than 11 years after its launch. TOMS provides daily global maps of the distribution of all the ozone in the atmosphere between it and the earth's surface. This quantity is called *total ozone*, and it is measured in units referred to as *Dobson units*, after G. M. B. Dobson, a British physicist in the early 20th century who devoted much of his career to the study of atmospheric ozone. The maps that TOMS provides of ozone concentrations are always depicted in Dobson units rather than parts per billion or million.

An instrument designed to look at ozone in the atmosphere in a different way was launched in February 1979. It is called SAGE (stratospheric aerosol and gas experiment), and as its name implies, it measures the distribution of aerosol particles and ozone in the stratosphere. The TOMS instrument looks down (what's called *nadir viewing*) and thus can't determine the distribution of ozone as a function of altitude. The SAGE instrument looks to the side toward the sun and measures the amount of light blocked out by the earth's atmosphere every time it encounters a sunrise or sunset—which an orbiting satellite does nearly every orbit. As the SAGE instrument locks on the track of the sun as it rises and sets, it obtains a vertical profile (i.e., a head count) of aerosol particles (dust) and ozone between the top of the atmosphere (somewhere below 200,000 feet) and the tops of clouds. On clear days with no clouds, the SAGE instrument data are reliable down to 15,000 feet. Although the SAGE instrument has an advantage over the TOMS because it can measure ozone as a function of altitude (primarily in the stratosphere), it obtains such profiles only twice per orbit, or approximately 30 times per day. Since it takes about 40 days for the SAGE to cover the entire earth, it can never achieve the one-day snapshot of the ozone distribution that TOMS is capable of providing. Furthermore, even though the SAGE measurements are uniformly spread across the earth's surface over this 40-day period, they are generally separated by a distance of about 1,000 miles.

The first SAGE instrument worked for 34 months, from February 1979 until November 1981. A second instrument, SAGE II, was launched in October 1984 (from the same space shuttle flight that obtained the second MAPS carbon-monoxide data-set discussed earlier) and is still providing measurements.

Scientists have found both SAGE and TOMS valuable tools in monitoring global ozone. The method used to take advantage of the satellite measurements is to take the distribution of ozone in the stratosphere obtained by SAGE and calculate the amount of ozone contained in a column of the atmosphere above the top of the troposphere. At the same time, scientists can look at the amount of total ozone from TOMS taken at the same location measured by SAGE and determine how much ozone is in both the stratosphere and troposphere. By subtracting the SAGE data from the stratosphere from the TOMS total ozone, scientists are left with an indirect measurement of the amount of ozone in the troposphere.

Figure 29 shows the distribution of this quantity, called the *tropospheric residual*. This particular figure is based on more than 8,000 data points obtained by SAGE between 1979 and 1987, and it shows the average distribution of the amount of ozone between the months of July and October. The predominant feature is the relatively large amount of ozone showing up west of the southwest coast of Africa. At this particular latitude the

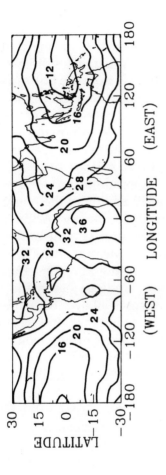

Figure 29. The distribution of tropospheric ozone in the tropics. This analysis was derived from two sets of satellite measurements between 1979 and 1987, and it depicts the average distribution during that time for the period between July and October, which corresponds to the "dry season" over southern Africa and eastern Brazil. The high levels of ozone found off the coast of Africa are believed to be the result of widespread biomass-burning, which is most pronounced during the dry season. The units on the contours are Dobson units and indicate the total amount of ozone in a column of air in the lowest 16 kilometers of the atmosphere.

wind generally blows from east to west, so what we see off the coast of Africa is a plume of ozone pollution that originated on the continent itself. Such a pattern, of ozone originating in one place and being carried by prevailing winds, is similar to the "ozone river" in the northeast corridor of the United States discussed earlier, and to the way smog travels from the Los Angeles basin eastward to Riverside in southern California.

In May 1989, at a conference on fire ecology held at the University of Freiburg, West Germany, scientists reported the most comprehensive studies ever undertaken on emissions from African fires. Their report stunned the scientific community and appeared on the front page of the June 19 edition of the *New York Times* with the headline "High Ozone and Acid Rain Levels Found over African Rain Forests." One of the scientists in charge of the study was Meinrat "Andi" Andreae, an atmospheric chemist from the Max Planck Institute in Mainz, West Germany. Andreae has made extensive measurements of trace gases all over the world, including over the Amazon basin in Brazil. But what he found over Africa made the Brazilian burning look like a backyard barbecue. As quoted by the *Times*, Andreae said, "We found the most serious pollution I have ever seen over a forest."

North of Brazzaville, the capital of the Congo, the research team flew 200 miles east and west, and according to Andreae, "there seemed to be no beginning or end to the pollution." Paul Crutzen, Andreae's colleague at the Max Planck Institute, reported at the same conference that because of their burning of vegetation, African fires were the world's leading contributor to pollution of this type. African savannah fires are so extensive, Crutzen said, that they pump three times more gases and particles into the air than all the fires set by farmers and settlers in South America, including the dramatic fires of the Amazon. Of all the carbon dioxide released by burning or deforestation, at least half of it comes from African fires.

"The forest burns once and it is destroyed," Crutzen was reported as saying. "But the savannahs and the grasslands become larger and are burned regularly. Shrublands used to be burned every three years. It has now become evident that they are burned every two years and even every year."

Unlike in the Amazon region, burning in this part of Africa goes on all year, although data from Brazzaville indicate that the highest ozone levels are found from August through October, a period that corresponds to the dry season over that part of Africa.

Another result of burning shrublands in Africa is acid rain. Scientists found the acidity of the rain over the Congo to have a pH of 4, or almost 10 times the normal acidic level. With respect to the acid rain problem, Andreae noted, "We also don't know yet how sensitive the tropical forest is to this [acid rain]. But with a mix like this in Germany or the U.S., we would have to look for damage in the forest."

Such a story points to a perplexing problem: Although the industrialized nations of the temperate zones are turning their attention to the sources of their own pollution, the nations in the poorer regions of the tropics can ill afford such a luxury. We've already discussed some of the problems facing this part of the world: the pollution from primitive agricultural practices, the rate of population growth, and the destruction of natural rain forests. This is not to cast blame for the global ozone crisis on the poorer nations that span the equatorial climate zone, but simply to emphasize that the ozone problem is global in nature—and that it is not so easy for the rich, industrialized nations to tell their poorer cousins that they can no longer practice agriculture as they have for centuries, or that they will not get a chance to develop because the already-developed nations have dirtied the place up so there's nothing left. The first step in solving a problem is to acknowledge the problem exists. What seems obvious is that despite national borders and differences among peoples, we are all residents of the same smoke-filled room—the planet earth—and that its continued existence is everyone's responsibility: rich nation and poor, white and black, temperate and tropical. Only with an all-out effort that spans national boundaries and local self-interests can we utilize the existing technology to save our planet.

CHAPTER TEN

The 1990s

The Decade We Did Something for Our Atmosphere?

In 1988 *Time* magazine's Man of the Year was Planet Earth. The theme of the issue was that the habitability and future of our planet had been the central focus of hundreds of important news stories written throughout the year. The cover of the December 1988 issue of *National Geographic* was a three-dimensional hologram of an exploding planet earth with this haunting question posed underneath: "Can Man Save This Fragile Planet?" The July 24, 1989, issue of *Newsweek* focused on problems of the environment and "cleaning up our mess."

So after nearly a decade of neglect, there was a new sense of commitment to cleaning up our environment. This renewed dedication has received support at both national and international levels, and within both political and scientific circles.

A FAVORABLE POLITICAL ARENA

During the 1970s, many of the new breed of conservatives who later helped propel Ronald Reagan into the presidency felt the Environmental Protection Agency was out to destroy big business. Reagan put the EPA under the control of Anne McGill Buford, who, according to Gregg Easterbrook, author of a feature article in *Newsweek*,[1] was "arguably among the least qualified individuals ever to hold an important political office." In his article, Easterbrook wrote further:

> Reagan staged several pitched battles against the pollution controls, notably a 1982 attempt to enervate the Clean Air Act. He suffered a loss of political capital when this offensive failed. Buford resigned in disgrace and her deputy, Rita Lavelle, was jailed for perjury. Several executives interviewed for this story described the failed assault on the Clean Air Act as what convinced them pollution control was here to stay: if Reagan at the crest of the conservative whitecap could not reverse the momentum for environmental control, then it could not be reversed.
>
> After 1982 Reagan never took on environmentalists directly again. But his OMB (Office of Management and Budget) continued to pocket-veto regulations by refusing the EPA permission to print them in the Federal Register. For the second half of Reagan's term, the EPA was run by Lee Thomas. Within the EPA, Thomas is admired for having fought a brilliant rear-guard campaign to keep the agency animate until the next administration.[2]

The political situation has changed under George Bush. His choice to head the EPA was William Reilly, a former head of the proenvironment Conservation Foundation. He is well respected and so far there are signs that the administration is providing the EPA with a lot

[1]*Newsweek*, July 24, 1989, pp. 26–42.
[2]Ibid.

of input rather than mere lip service. Hopefully, this is an indication that George Bush is taking his campaign promise of being an "environmentalist" to heart. In addition to presidential support, the leadership in the Senate has passed from Robert Byrd of West Virginia to George Mitchell, a citizen of Maine. Byrd was soft on pollution since his state was so dependent on the production of soft coal and petrochemicals. Mitchell, on the other hand, is an environmentalist from a nonindustrialized state that is being soaked with acid rain.

Thus the political stage appears to be set properly to achieve new goals to clean up the air in the present decade. An even more encouraging fact is that the international community is finally recognizing the need to band together to save our environment.

THE FIGHT FOR A REAL CLEAN AIR ACT

The legislation passed in the 1970s directed at the abatement of ozone pollution in the United States badly missed the mark. In trying to solve the global pollution problem it did virtually nothing toward controlling the emissions of nitrogen oxides. One reason for this disastrous omission was the lack of scientific understanding at the time about how ozone is formed. Although hydrocarbon emission reductions were important for controlling the severe pollution problems characteristic of Los Angeles in the 1950s and 1960s, new research has shown that such controls have had virtually no impact on the broader ozone-pollution problem. Hopefully, such shortcomings in our scientific knowledge are past history, like the flat-earth theory and the pre-Copernican cosmology that held that earth was the center of the universe. But only time will tell.

On June 12, 1989, President Bush delivered a speech to the state governors' conference which sounded as though he was ready to "take the bull by the horns" to finally solve our national pollution problem. In his speech he told the governors:

> Too many Americans continue to breathe dirty air, and political paralysis has plagued further progress against air pollution. We have to break this log jam by applying more than just federal leverage. We must take advantage of the innovation, energy, and ingenuity of every American. The environmental movement has a long history here in this country. It's been a force for good, for a safer, healthier America. And as a people, we want and need that responsibility and respect for the natural world. And this will demand a national sense of commitment, a new ethic of conservation, and I reject the notion that sound ecology and a strong economy are mutually exclusive.

The President delivered his proposed legislation to Congress a month later. The plan called for the first sweeping revisions in the Clean Air Act since 1977. In his proposal, the President focused on curbing three major threats to the environment: acid rain, urban air pollution, and toxic air emissions. The proposal, in the context of the previous decade when the environment was completely ignored by the administrative branch of government, had some admirable goals. But in the area of urban air pollution most of the President's proposal focused on the reduction of volatile organic compounds (VOCs). The proposal gives the EPA authority to regulate VOC emissions from small sources and consumer products. Also in the plan is a tightening of standards pertaining to hydrocarbon emissions from cars and trucks, along with expanded vehicle inspection and maintenance programs in areas which are "nonattainment" areas (i.e., those localities not in compliance by the deadline).

President Bush's proposal was greeted positively by his supporters, but not all shared this view. Senator Mitchell responded by saying "the proposal contains much that deserves praise, primarily the simple fact that it was made at all. But, on balance, I am disappointed. The legislative proposal is weak in comparison to the strong statements by the President on June 12."

Mitchell, who had already introduced a bill in the Senate a year earlier, was especially critical of what he considered President Bush's method of reasoning. According to Mitchell, whenever a tough decision concerning the environment and economics needed to be made, the President's choice seemed to favor economics every time. This was particularly true in the case of regulation of nitrogen oxides. Mitchell stated that the President's proposal did not require net reductions in NOx emissions under either the acid-rain or the urban-air-quality sections. "Incredibly," Mitchell said, "the President proposes to permit increases in NOx emissions between now and the year 2000. This is unfortunate because NOx is a key component in the formation of both ozone and acid rain."

If the President's proposed legislation is passed by Congress, ozone pollution will most likely become worse, not better. Mitchell's proposed legislation, as well as some other suggestions currently before the House of Representatives, mandates restrictions on tail pipe emissions of nitrogen oxides. The state of California has already put new pollution standards into effect that include the control of NOx emissions from automobiles. These provisions must be incorporated into federal legislation if the widespread smog problem is ever to be solved. There will be considerable opposition in Congress, especially from those members whose constituents are dependent upon the health of the automobile industry. It will be of crucial significance what type of legislation Congress finally passes and whether or not the President comes through with his signature if the legislation requires strict regulation of nitrogen oxide emissions.

THE MONTREAL PROTOCOL: COMBATING CHLOROFLUOROCARBONS

On September 16, 1987, the Montreal Protocol was signed by 24 countries, including all of the principal producers of chlorofluorocarbons (CFCs). This document requires all signatory countries to reduce their consumption of CFCs by 20 percent (relative to 1986 usage) by 1993, and by 50 percent by 1998. The Montreal Protocol marks a significant achievement in solving a global problem. But of the three components of global change (depletion of the ozone layer in the stratosphere, greenhouse warming, and global smog) that we have discussed, depletion of the ozone layer in the stratosphere is the one that should be the easiest to control, since it is the only one of the three problems that involves a chemical entirely made by humans.

There would be no CFCs on this planet if human beings were not here.

On the other hand, carbon dioxide and ozone are integral components of life on earth. The problem we face with these two gases is that unnatural concentrations of them may exist as a direct result of human activity. Increased amounts of carbon dioxide might have been considered harmless except for the fact that sophisticated theoretical calculations suggested otherwise. We now know that too much carbon dioxide will eventually warm our planet. The direct harm to earth as a result of higher levels of ozone is an even more severe problem, since environmental studies are just now beginning to show that too much ozone in the air may have detrimental effects on human health as well as on our forests and crops.

Why has it taken us so long to see the problem? One reason is that in science, as in other fields, there is a vast gray area of knowledge, where facts and hypothoses exist in murky confluence. One person's conclusive proof is another's theory; for every fact that surfaces there is a waiting barrage of tests and doubts. For nearly every example of something negative that has happened to the environment, another example can be found where, for some unknown reason, no harm resulted. Ozone, for example, when first discovered, was thought to be a cure-all for human ailments. Many doctors prescribed "ozone treatments" for a variety of illnesses.

Back in 1974, the impact of CFCs on the ozone layer was likewise only a theory, a hypothesis put forth by two atmospheric chemists: F. Sherwood Rowland and Mario

Molina, from the University of California at Irvine. They claimed that chlorine released from these manufactured chemicals would cause irreparable damage to the earth's ozone layer, the primary shield against harmful ultraviolet radiation. The theory was not universally accepted in the 1970s. There were too many "logical" questions that demanded to be answered:

—*Aren't there natural sources of chlorine in the stratosphere?* Careful measurements in the 1980s have shown that most of the chlorine in the stratosphere is manufactured.

—*Aren't there surely other reactions in the stratosphere that counteract the ones that cause chlorine to destroy ozone?* Here the doubters proved correct at first glance. Research in the chemical laboratories around the world has shown that new compounds such as chlorine nitrate and hypochlorous acid do exactly that: counteract the effect of CFCs. But these gases are not made quickly enough in the stratosphere to negate the detrimental impact of chlorine atoms on the ozone layer; they only slow the destruction rate.

When scientists began looking for trends in the stratospheric ozone measurements using data from the 1960s and 1970s, none were found. Some analyses even showed that ozone in the stratosphere was *increasing* during this time period. *If there was a loss of ozone due to the CFCs' releasing chlorine, then why didn't it show up in the data?* Perhaps there was a natural variability related to the 11-year sunspot cycle. Perhaps the late 1960s and the 1970s had seen a refurbishing of the ozone layer after some of it had been destroyed by atmospheric nuclear-bomb-testing in the 1950s and early 1960s. These were all questions that had to be answered before restrictions against CFCs could be imposed. No solution could be implemented to solve a problem that couldn't be proved to exist.

But in the late 1970s and early 1980s, the amount of ozone in the stratosphere measured by satellites started to go down. *Was this decrease part of the natural cycle? Or were the satellite instruments at fault, since they weren't designed to measure trends?* Another group pointed to the spate of recent volcanic eruptions—Mt. St. Helens, El Chichón, and Ruiz—that had occurred in the early 1980s and questioned the data from the satellites because of all the dust released into the atmosphere, which may have caused the instruments to provide erroneous data.

So the timing was right in 1985 to set up a blue-ribbon panel to try to answer all the doubters, to focus on the questions about the apparent decrease in ozone levels in the stratosphere. This panel was convened just at the right time. For meanwhile, in Antarctica, the "smoking gun" had been found—proof positive that yes, indeed, ozone levels were decreasing in the stratosphere. The smoking gun, of course, was the "ozone hole" over Antarctica, and had it not been for the publicity given to that discovery, the exhaustive study undertaken by the distinguished blue-ribbon panel (which received input from over 100 scientists from around the world) may not have stirred much interest.

In conjunction with the panel, in 1987 an international expedition was mounted to measure the important chemical constituents related to the apparent hole

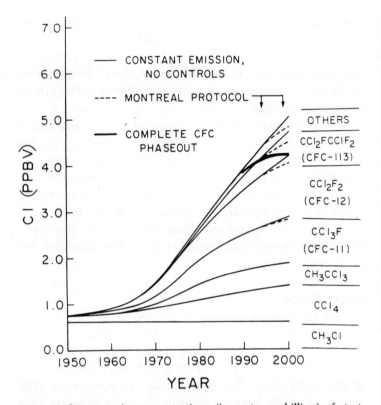

Figure 30. Increases in concentrations (in parts per billion) of stratospheric chlorinated molecules. The solid curves assume continued release of chlorofluorocarbons at 1986 rates. The dashed curves assume a 20 percent reduction in release rates in 1994 and an additional reduction of 30 percent by 1999, a scenario consistent with the Montreal Protocol. The heavy solid line assumes a complete phaseout of chlorofluorocarbon emissions over a 10-year period starting in 1989. Each of these molecules is present in the atmosphere because of industrial processes, with the exception of methyl chloride (CH_3Cl), the molecule at the bottom of the list, which has a natural atmospheric concentration of 0.6 parts per billion by volume (ppbv). The primary source of methyl chloride is photosynthetic

in the ozone layer over Antarctica. The results of that expedition pointed conclusively to CFCs as the main culprit for the decreasing ozone. Headed by NASA's Bob Watson, the trends report showed convincingly that even when all the other factors, such as volcanoes, sunspot cycles, and satellite instrument drift, were factored in, ozone levels in the stratosphere were indeed decreasing—not only over Antarctica, but everywhere in the stratosphere.

The Montreal Protocol, then, was a final step, the culmination of years of research. Rowland and Molina had been correct from the start. Yet the world had not followed their calls for immediate action, which had been made almost a decade earlier, until the damage was already done. As Figure 30 shows, the gradual phase-out resulting from the Montreal Protocol will have a minimal impact on the amount of chlorine in the stratosphere by the beginning of the 21st century.

The major question asked by most people is how much their lives will really change as a result of the Montreal Protocol. Probably not very much, if at all. Already chlorofluorocarbons have been phased out of use by manufacturers of spray cans and aerosols. They managed to find suitable replacement for the offensive ingredient without much added expense.

---

activity taking place in the oceans. Even the implementation of the Montreal Protocol, which will reduce emissions of anthropogenic halocarbons, will have little impact on the chlorine content of the atmosphere over the next several decades. (Figure from F. S. Rowland, "The Role of Halocarbons in Stratospheric Ozone Depletion," in *Ozone Depletion, Greenhouse Gases, and Climate Change.* Washington: National Academy Press, 1989, pp. 33–47. Reprinted with permission.)

On the other hand, many of us would miss the use of our air conditioners if we were suddenly told that we couldn't use them. But that's not going to happen, either. In the United States, one of the largest sources of CFCs released to the atmosphere is automobile air-conditioning systems. Years ago, chemical manufacturers began looking for a substitute, a gas that could be used in air-conditioning systems that wouldn't eat up stratospheric ozone. They found it in fluorocarbon-134a (FC-134a). Its chemical properties are much like those of some of the chlorofluorocarbons, but it doesn't have the *chloro* in front of it. The fluorocarbon-134a molecule resembles the ethane molecule discussed in Chapter 3, but instead of the six hydrogen atoms usually found in ethane molecules, four have been replaced by fluorine atoms. And three of the fluorine atoms are attached to one carbon atom, while one fluorine atom is attached to the other carbon atom. This arrangement makes the molecule assymetrical (thus the *a* in *FC-134a*). When FC-134a is released into the atmosphere it cannot release any chlorine when it gets broken down because there isn't any chlorine. Furthermore, because FC-134a contains hydrogen atoms, it can react with hydroxyl (OH) radicals, thus preventing FC-134a from accumulating in the atmosphere.

So the American consumer doesn't have to sacrifice too much to save the ozone layer in the stratosphere. FC-134a can be substituted for CFCs in just about every air-conditioning system in use today. Manufacturers are already beginning to build pilot plants to start making the gas, and it should reach the market by 1993. So the goal of the Montreal Protocol of a 50 percent reduction in CFCs by 1998 is easily attainable.

CONTROLLING THE GREENHOUSE EFFECT

Unfortunately, dealing with the problem of the increasing levels of carbon dioxide (CO_2) requires a more complicated solution than that for CFCs. For decades we've been warned of global warming due to increases in carbon dioxide concentrations. We know that the primary cause of this increase is our use of fossil fuels. We know that deforestation on a global scale is contributing to carbon dioxide increases. We know that there are other trace gases that likewise contribute to greenhouse warming, such as methane, the chlorofluorocarbons, and tropospheric ozone. In addition, the world's population continues to increase, especially in the tropical areas, and with this increased population come increased levels of methane. Two of the primary sources of methane are rice paddies, because of the standing water which promotes the growth of microorganisms that make the gas (analogous to what is called *swamp gas* in the U.S.), and cattle, whose enteric digestive system generates large amounts of methane. Both rice and cattle production will increase along with human population. So as the world's population grows, the quest to control the increase of carbon dioxide and methane becomes an even greater challenge.

If the Montreal Protocol works, not only will our air be cleaner, but an important step toward controlling the greenhouse effect will have been taken. By eventually phasing out the use of all chlorofluorocarbons, we can get rid of a gas that contributes an estimated 10 percent to global warming. But even without chlorofluorocarbons, we are still left with carbon dioxide. Scientists estimate that human activity puts about 7.4 billion tons of carbon dioxide into the atmosphere each year from burning fossil fuels and wood. They claim that we can head off the warming due to increased carbon dioxide if we cut emissions by two-thirds. Replanting forests can recycle as much as 2 billion tons of carbon dioxide per year. Within a few decades we would have to plant an area equal to 10 times that of the state of Oregon in order to have such a significant effect on CO_2.

Replanted forests, however, are not a permanent sink for carbon dioxide. After several more decades, the CO_2 stored in the forests will be released into the atmosphere again when the trees die and decompose, or if they are burned at some future date. But reforestation will stem the tide, even if only temporarily, giving us a reprieve until well into the 21st century. The cost of such a massive replanting on a global scale is estimated to be about $200 billion. And if, while we plant trees, we can cut the rate of deforestation in half, scientists estimate we'd reduce carbon dioxide emissions by another 1.1 billion tons per year.

Here's another thing: The rate of population growth in the undeveloped countries is three times faster than in the so-called industrialized countries. If the rate of growth of these nations can be cut in half, scientists estimate that the rate of CO_2 emissions will be cut by another 700 million tons.

And yet another option is the use of nuclear energy. This is becoming more and more unpopular in the United States, in large part because of the few cases such as Three Mile Island and Chernobyl where nuclear plants have had major accidents. Unfortunately, the response to such accidents has obscured the simple fact that *if* it could be developed safely, and *if* something could be done about nuclear waste, nuclear energy is a wise alternative which would succeed in reducing carbon dioxide emissions. Currently France generates most of its electricity with nuclear power, and the safety record of the generating plants is excellent. By shifting our power generation to nuclear rather than fossil fuel, scientists estimate we could reduce carbon dioxide emissions another 700 million tons.

In some areas nuclear power might not be the best alternative. Renewable sources of energy such as the sun, the wind, and hot springs have so far proved too costly to develop on a large scale. But shifting from fossil fuel to these sources, where appropriate, would reduce carbon dioxide emissions another 700 million tons. One means of paying for such development was suggested, in congressional hearings on the subject, by Stephen Schneider, atmospheric scientist at NCAR in Boulder, Colorado. He proposed a tax on fossil fuels to be used to subsidize the development of alternative fuels for generating power.

Besides developing alternative sources of power, another step we can take is to use the fuel we have more efficiently. Building automobiles that use less gasoline is one step. Insulating our homes so we can heat and cool them more efficiently is another. On a global basis, if the industrialized nations, which consume more fuel than the less developed nations, can use fuel twice as efficiently as in the past, scientists estimate they can reduce carbon dioxide emissions by 1.5 billion tons.

Individually, these steps seem small, but collectively they add up. By taking all the above steps we could reduce carbon dioxide emissions by the two-thirds necessary to head off the greenhouse effect. The cost of such an effort is estimated to be about $150 billion per year. Spread among the industrialized nations of the world, that amounts to about $100 per year per person. We already spend six times that on national defense in the United States. What's more important? What should be the priority?

WHO'S GOING TO PAY FOR IT?

During 1987 and 1988, the topic of the earth's environment received unprecedented attention by Congress, which held nearly two dozen congressional committee hearings on matters pertaining to global change research. In the first four months of the 101st Congress, 10 bills were introduced on the subject. Despite all this flurry of activity, relatively little has actually been done. As of 1989, the latest global change legislation passed was Senator Joseph Biden's Global Climate Protection Act of 1987, which was passed as part of the 1988 Department of State authorization bill (P.L. 100-204). The main thrust of the bill was to identify which agency is responsible for the enactment of any legislation pertaining to global change. It divides the responsibility between the Environmental Protection Agency and the Department of State. Before the end of 1990, both agencies are to report back to Congress on federal scientific and policy efforts to deal with the greenhouse effect and the international effort to address the overall problem of climate change.

A more positive note is that agencies such as NASA, the EPA, the National Oceanic and Atmospheric Administration, and the Department of Energy, and the

National Science Foundation have included budgets for conducting further research on global change. One of NASA's major efforts during the 1990s will be the deployment of the Earth Observing System (EOS), a joint international effort with the European Space Agency and the Japanese Space Program to study the earth's oceans, atmosphere, and geology. The participating countries plan to launch four platforms, two from the U.S. and one each from the other two agencies, by the beginning of the next century. From these platforms scientists will undertake further research on the earth's atmosphere. The first platform is due to be launched in 1998. But the program will cost. In testimony before Congress, Frank Press, president of the National Academy of Sciences, said, "Clearly an effort of this scope and magnitude presents a rather awesome challenge to science and to the institution of science. It presents, as well, a challenge to our political system."

In the Spring 1989 issue of *Earthquest* magazine, Thomas R. Kitsos and Daniel M. Ashe, two congressional staff members, wrote:

> Whether the political system and the scientific community are up to the challenge remains to be seen. To get the system moving, however, scientists must overcome their almost instinctive reluctance to get involved in politics. The socialization of a scientist seems to breed a type of caution that is essential in research but frequently dysfunctional in the political world. Given the fiscal environment in which the nation finds itself today, and the need of the U.S. Global Change Research Program for additional federal funds, this is not a time for timidity. Eventual world habitability may depend on a political aggressiveness heretofore unheard of from the scientific community.

The plans that NASA has for its Mission to Planet Earth are impressive, but the "big bucks" to make the Earth Observing System, the centerpiece of the program, a reality still have not been committed. The feasibility studies and preliminary plans done so far have been relatively inexpensive. Detailed plans have already been formulated for the instruments that may eventually fly on EOS. Once the platform is designed and the instruments are selected, the nation's commitment to NASA's global change program will add up to billions of dollars. By the mid 1990s, it will be evident whether or not our country is willing to put forth the money to support the global change research in the way scientists have been urging.

THE THREAT OF TROPOSPHERIC OZONE

One of the goals of this book has been to recognize the trilogy of global change: depletion of the ozone layer in the stratosphere, climate warming due to the buildup of carbon dioxide, and the increasing concentrations of ozone in the lower atmosphere. Ozone pollution has traditionally been identified as a local or regional problem. But new findings have shown that ozone pollution is a threat not only to those of us who live in or near urban centers. It is now a problem that threatens our forests, our ability to produce food, and our health. Living with higher ozone levels is the most serious aspect of the trilogy of global change.

To add to the problem, there appears to be a synergistic effect between the other two parts of the trilogy and tropospheric ozone. As the ozone in the stratosphere is depleted, for example, more ultraviolet radiation reaches the earth, thus accelerating the photochemical process that produces ozone. The commitment of the Montreal Protocol to stopping the thinning of the ozone layer will have a positive but not a major impact on the global increase of tropospheric ozone concentrations.

The increasing accumulation of both carbon dioxide and global smog has its roots in the burning of fossil fuels and biomass. If we are ever to implement a plan to reduce the buildup of carbon dioxide in the atmosphere, then the use of smaller amounts and more efficient methods of combustion for our energy needs will eventually have a considerable impact on the return of tropospheric ozone concentrations to their natural levels.

But the tropospheric ozone problem must be addressed with much greater haste than the carbon dioxide problem. Tropospheric ozone is increasing two to three times faster than carbon dioxide in the northern hemisphere. However, there is more at stake than just global warming, although tropospheric ozone may be contributing to that problem even more than increasing levels of carbon dioxide. Ozone is a toxic gas. As it increases in the atmosphere, it will kill forests and crops, and it will cause major damage to our lungs. The need to abate its buildup is urgent. Decades may prove to be too late as we watch our forests die and are faced with unprecedented epidemics of respiratory disease.

For whatever good it may do, the ozone problem, in this country and much of the industrialized world, is also seen as a pressing local pollution problem. Thus legislation that curbs local ozone pollution is a necessary first step in reducing the impact of the third and deadliest component of global change.

COMMITMENT TO BETTER SCIENTIFIC
UNDERSTANDING

In order to fully appreciate the extent of the global ozone-pollution problem and the complex reasons for the formation of ozone on a worldwide scale, the cooperation of the international scientific community is required. A significant step was taken in November 1988, when an international workshop attended by some 50 scientists from 13 nations was convened in Dookie, Australia, to develop a plan to study the earth's atmospheric chemistry on a truly global scale during the 1990s. The International Atmospheric Chemistry Program was created in response to the growing concern about the rapid atmospheric chemical changes and their potential impact on human civilization. The overall goal of the program, as stated in the Executive Summary of the Dookie workshop, "is to measure, understand, and thereby predict changes over the next century in the chemistry of the global atmosphere with particular emphasis on changes affecting the oxidizing capacity of the atmosphere, the impact on climate, and the atmospheric chemical interaction with the biota." Note that the first part of the goal is devoted to understanding the "oxidizing capacity of the atmosphere," which is closely related to understanding the formation and the resultant distribution of tropospheric ozone.

The Dookie workshop goal is broad and encompasses several urgent environmental issues, including the increasing acidity of rainfall, the depletion of stratospheric ozone, the greenhouse warming due to trace gas accumulation, and the biological damage from increased ozone levels. One of the underlying themes in the plan of action is that the scientific community still does not completely understand the natural variability of the earth's chemical system. Any future action by society to control its altering of the composition of the atmosphere must take into account quantitatively how the makeup of the atmosphere would have changed naturally had humankind not been present. Without such knowledge of the natural state, and hence the natural variability, of the stratosphere, for example, the Montreal Protocol never would have become a reality.

In addition to the specific scientific goals outlined by the Dookie workshop participants, the need to educate the public was listed as one of the primary focuses. It is important for each of us to understand that local pollution and, in particular, high levels of ozone have had important repercussions that have contributed collectively to a global problem. This awareness seems to be happening now. In the fall of 1989, for example, eight northeast states (the six New England states plus New York and New Jersey) agreed to require emission controls on automobiles in their states to adhere to the stricter California standards, which reduce nitrogen oxide emissions. One of the state officials said that such an action was sending a message to the federal government that the citizens of those states would not drag their feet in working toward cleaning up their environment. Such action needs to be repeated, for only when each individual commits himself or herself to cleaning up the environment can we truly stem global smog.

Epilogue

∿∿∿∿∿∿∿∿∿∿∿∿∿∿∿∿∿∿∿∿∿∿∿∿

The year is 2004. Sam and Melissa are standing in a crowd with other graduates of the University of Maine. It is early June in New England, but already the season's first hot spell has settled over the area. As the two wait for their diplomas and look toward their future, the prognosis is more hopeful than it was 18 years before when they were first entering the school system. There are still problems, but there have also been changes. The tide has turned, and for them and the rest of the students about to embark on their lives and careers, the future looks brighter than it did for the previous generation.

There are several reasons for this: The Earth Observing System (EOS) has been in place for several years, and already the data coming out of the program have proved valuable. For example, the EOS data have shown that the ozone "hole" over Antarctica has stopped growing as fast as it was in the late 1980s. At the same time, instruments on board the EOS have shown that the actual amount of chlorine in the stratosphere remains fairly constant. As fewer and fewer chlorofluorocarbons have been released into the atmosphere, the ozone in the stratosphere should start replenishing itself. Yes, the earth is getting warmer, but thanks to the watchful eye of the world's scientists and leaders, those affected by the warming will be well prepared. For example, the EOS data show that the Antarctic ice layer is thinning, a sign of global warming. Already the governments around the world are beginning the process of preparing for the changes. In Bangladesh, the government, with help from the industrialized nations, is encouraging residents along the low-lying coast of the Bay of Bengal to move inland to higher ground as the water already is intruding on the shoreline. In the United States, certain areas of the Gulf Coast and Florida's Atlantic coast are preparing in a similar manner.

There are some things not available to Sam and Melissa's generation that were available to their parents and grandparents. The automobiles they drive, for example, are smaller and lighter than those of a generation ago. Much of the metal in the car bodies has been replaced by stronger and lighter composite materials. Some new models even have engines constructed entirely from nonmetallic material. They have emission control systems that have virtually eliminated the release of nitrogen oxides into the atmosphere. Speed limits have been lowered to better conserve fuel, and there are now appearing in gas stations alternative fuels such as gasahol, which combines gasoline and alcohol, which saves more money and allows cleaner combustion.

But that's not all. Sam and Melissa live in a country, the United States, which has finally awakened to the need for mass transportation. In cities from Boston to Los Angeles, the past decade has seen a spurt of new construction in the inner cities, the development of fast, clean electric rail lines or the refurbishing of existing mass-transit systems. The federal government has undertaken a comprehensive plan to link the country via railroads, upgrading the tracks and service of Amtrak. And as Sam and Melissa step out into the world, they will find their lives different from those of their parents and grandparents in at least one profound way: cars will no longer be a status symbol. It is "in" to ride the train to work, cooler to hop aboard the train with your bicycle and backpack than to load the RV with electric gadgets and "see the USA." You can still see the USA, but a new breed of Americans enjoy the more leisurely method of highway travel as they watch the countryside's forests rebound from the bombardment of high ozone levels during the 1990s.

Sam and Melissa's goals are also different. With the global smog problem now in control, the world has been shocked out of its complacency. People are aware of the environment in a way undreamed of 100 years previously. Sam has a degree in city planning and will be working as a volunteer in the newly established Youth Volunteer Corps, helping build the light rail system for Los Angeles. Melissa is a chemical engineer and will do her year of volunteer service in Burkina Faso, helping farmers find alternatives to burning fields.

As Sam and Melissa look toward their future, they are perhaps the first generation in America in a long time who are aware in a fundamental way of their place in the larger world. Recycling is a way of life now. Landfills are becoming so crowded that most towns and cities charge residents to dump their trash; it is more economical for them to recycle all they can. The recycling business—redemption centers, crusher plants, brokers—has become the boom business of the decade. The generation of graduates entering the world at this time know what can be done and are determined to do it. The doomsayers of the late 20th century have been proved wrong, at least so far.

Such a scenario is only speculation, but it can happen. The technology is already known that can rectify many of the ills that have led us down the road to global smog. Whether we decide to turn the tide or to continue needlessly wasting our resources is up to us. Whether to leave our children our filthy mess or to start cleaning it up is something each individual can decide right now. The future is in our hands.

Index